THE PROOF
AND
THE PUDDING

Jim Henle

THE PROOF
AND
THE PUDDING

What Mathematicians, Cooks, and You Have in Common

Princeton University Press

Princeton and Oxford

Published by Princeton University Press, 41 William Street,
Princeton, New Jersey 08540
In the United Kingdom: Princeton University Press, 6 Oxford
Street, Woodstock, Oxfordshire OX20 1TW
press.princeton.edu
Jacket illustration by Leon Steinmetz
Cartoons by Leon Steinmetz
Library of Congress Cataloging-in-Publication Data
Henle, James M.
The proof and the pudding : what mathematicians,
cooks, and you have in common / Jim Henle.
pages cm
Includes bibliographical references and index.
ISBN 978-0-691-16486-1 (hardback : alk. paper) 1. Mathematical
recreations. 2. Mathematics—Miscellanea. 3. Cooking—
Miscellanea. 4. Food—Miscellanea. I. Title.
QA99.H46 2015
510—dc23
2014049572
British Library Cataloging-in-Publication Data is available
This book has been composed in Baskerville 10 Pro
Printed on acid-free paper. ∞
Printed in Canada
1 3 5 7 9 10 8 6 4 2

CONTENTS

PREFACE

THE PREMISE of this book is that if you look at mathematics and gastronomy the right way, they are amazingly alike. The goal of this book is to explore the two and to reveal their essential similarity.

But of course, math and cooking aren't the same. You don't eat numbers. You can't take the square root of a muffin. And for most diners, a real bowl of spaghetti is more satisfying than any number of theoretical feasts. But that's the wrong attitude.

Instead, we're going to think about what cooking and math do have in common: They began as crafts, simple and useful. They evolved into arts, complex and pleasurable. Both are presented, often, as lists of instructions. Both feature taste. Both pose difficulties. Both celebrate champions. Both intimidate novices.

That's just the start.

But you must be warned. This book is about mathematics and cooking, but it won't draw any connection between them. In particular, *there are no applications of math to cooking*. Math is useful in the kitchen, sometimes, but that's not what this book is about.

Furthermore, mathematics itself isn't the point of this book. I'll show you some math, really cool math (incredible stuff, actually), but the purpose will be to illustrate features of mathematics. This isn't a "math" book.

Food, also, is not the point. There are recipes, good recipes, frighteningly good recipes (you have no idea). But they serve a higher purpose. This isn't a cookbook.

After reading *The Proof and the Pudding*, you *may* know more math. And you *may* know more about cooking. But that would be incidental. The goal is for you to see mathematics and gastronomy in a new way, as fraternal twins.

I intend to convince you that

1. We do math and we cook for more or less the same reasons.
2. We solve math problems with the same attitudes and tools we use to solve problems in the kitchen.

3. We judge mathematics with many of the same criteria we use to judge food.
4. And in general, life in mathematics and life in gastronomy are remarkably similar. Mathematicians and cooks have similar dreams, similar fears, and similar guilty secrets.

There's more that I want to say about cooking and math. I have gastronomic and mathematical views on aesthetics, creativity, inspiration, strategy, genius, and depravity. I thought I'd put them into one book.

And one more thing. This book has a deeper purpose. Underneath the propositions and the pastry, the real subject of this book is *fun*.

There are reasons to do mathematics and there are reasons to cook. But in both fields, the chief motivation is pleasure. This is not hard to see on the gastronomic side, but mathematicians also, surprisingly, are hedonists. They're out for a good time.

I don't want to sound immodest, but I think I'm good at having fun. You can ask my friends. They'll all tell you I have more fun than I should.

I can find pleasure in repetitive tasks. I can enjoy hopeless quests. I can have fun making serious mistakes.

If there's something I'm interested in, I will pursue it to excess. I will spend happy hours covering sheets of papers with straight lines and covering sheets of dough with pats of butter.

I'm not sure what this adds up to, but there's a zen to it and I have it. And I'd like to share it.

ACKNOWLEDGMENTS

I AM indebted to so many.

There are those who helped me with proofs. There are those who helped me with puddings. There are those whose support for the project and the ideas were crucial to its realization. There are members of my family who walk these pages, seen and unseen. Finally, as the poet says, they also serve who only sit and eat.

Most especially I want to thank Bill Zwicker and Cathy Brodie, David and Doris Cohen, Cutberto and Yolanda Garza, John and Matt Thorne, Carolyn Cox and Sam Perkins, Steve Spitz and Cynthia Ingols, Marjorie Senechal and Stan Sherer, Ron and Del Blank, Klaus Peters, Vickie Kearn, a generation of students, and Henles Allison, Fred, Portia, and Theda.

At the finish line I have been buoyed by the care, understanding, and enthusiasm of the staff at Princeton University Press—Dimitri Karetnikov, Carole Schwager, Mark Bellis, and though I mentioned her already, Vickie Kearn.

Some of the material in this book has been adapted from the Mathematical Intelligencer and permission has been granted by Springer Verlag.

Finally, the illustrations by Leon Steinmetz are quite amazing.

Jim Henle
July 2014

THE PROOF
AND
THE PUDDING

1

THE MAD SCIENTIST

WE'RE GOING to start with two investigations, one in mathematics and one in cooking. On the surface, they have little in common. There are similarities, though, similarities of "spirit," for want of a better word. I'll say more at the end.

DOODLES

A few years ago, I was doodling, that is, I was sitting with pen and paper, drawing lines with no purpose in mind. I drew a square with grid lines.

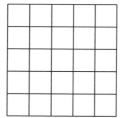

I put diagonals in some of the boxes.

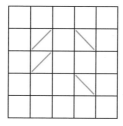

I imagined the lines as mirrors. I wondered what would happen if a ray of light entered the square and started bouncing around.

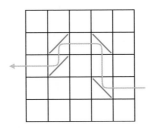

I noticed that it was possible for a ray of light to visit the same box twice, that is, it could bounce twice off the same mirror.

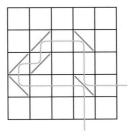

That made me wonder how long I could keep things going. Given a square, how long a path could I make, assuming I could place the mirrors wherever I wanted?

I started small. The 2 × 2 square allowed a path of length 5.

(I count each square I enter. I count a square twice if I enter it twice.)

For the 3 × 3 square, I first got a path of length 9,

then one of length 10,

then one of length 11.

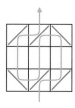

That seemed the best I could do.

Do? Do? What was I *doing*?

I was just having fun. First there was a grid. Then there were mirrors. Without a plan, I found myself drawing lines, scratching in diagonals, and tracing beams of light.

I was curious. I wanted to know how long a path I could make. I wanted to know the longest path in a 4 × 4 square, the longest path in a 5 × 5 square, and so on. After a while I got curious about rectangles, too.

You should understand that you're dealing with someone who gets very excited about primitive pen-and-paper activities. When I first thought of mirrors in a square, I drew grid after grid after grid after grid after grid.

The best I could do for the 4 × 4 was a path of length 22.

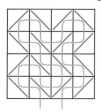

The best I could do for the 5 × 5 was a path of length 35.

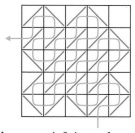

But that looks a little unsatisfying, doesn't it? There's a square (upper right-hand corner) I never visit. If I visited every square, could I get a longer path, couldn't I? But I've tried. I don't think I can!

After some thought, I was able to prove that my answer for the 4 × 4 square was the best that was possible.

You may be wondering: "Is this really mathematics?"

It is indeed mathematics. I take an expansive view of the subject. For me, any structure that can be described completely and unambiguously is a mathematical structure. And any statement about that structure that can be proved beyond doubt is a mathematical statement. The proof is a mathematical achievement. Inventing such a structure, making discoveries about it, proving statements about it—that's mathematics.

The structure of the square, the mirrors, and the rays of light can be described completely and unambiguously. And the fact that the longest possible path on a 4 × 4 is 22 is a genuine mathematical statement.

Here's my proof that the longest path on a 4 × 4 is 22:

Since we have an example of a path of length 22, all we have to do is show that no longer path is possible.

Now it's clear that we can visit a square at most twice, like this—

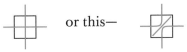

or this—

But we can visit an edge square only once,

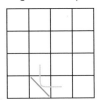

unless we are entering or exiting,

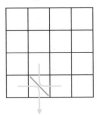

so the best we can do is

1. visit the four interior squares twice,

2. visit the edge squares once, except

3. visit two edge squares twice.

That makes a total of 22, as in the case of the example earlier.

1	1	1	1
2	2	2	1
1	2	2	1
1	1	2	1

And that's a proof that 22 is the best we (or anyone) can do.

I call this a "doodle." There are more doodles. Anyone can invent a doodle—you just invent your own rules. I've had fun, for example, with a one-way mirror doodle. It looks like this.

In one direction the mirror reflects and in the other it doesn't. Here's a little example.

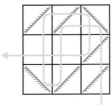

This can be a lot of fun. I think I can get a path 33 squares long in a 4 × 4 square.

There are doodles and doodles. I've set up a website, with notes for most chapters in this book:

press.princeton.edu/titles/10436.html

I invite the reader to visit the site. In particular, there are more doodles there.

I also invite you to share your doodle ideas with me.

jhenle@smith.edu

NOODLES

This may seem a *little* abrupt. Noodles and doodles appear to have almost nothing in common. I'll say something at the end of the chapter.

It started a few years ago. I was preparing a dinner for friends. I had planned to cook some sort of spaghetti dish.

But my plans were confounded when I learned that one of the guests had celiac disease, an allergy to wheat gluten. She couldn't eat wheat pasta.

That was bad news. But I was determined to cook that dish. At the grocery store I found corn spaghetti. I bought a box of it. I cooked it—and served my puzzled friends a gooey mess.

I *may* have overcooked it. But I suspect that the only way to undercook corn spaghetti is to leave it at the store.

My guest was embarrassingly grateful. I could have left it there but I saw a challenge that intrigued me. Is there something to take the place of pasta? Could I find a substance that

- is gluten-free, and
- is functionally equivalent to pasta?

No dumplings. No gnocchi. No couscous. All of these contain wheat.

What I wanted was a sort of multipurpose faux pasta, something that could comfortably take the place of macaroni or penne. I tried practically everything. French fries. Brussels sprouts. Corn flakes.

The challenge may seem difficult and maybe even pointless. But it attracted me.

And it entertained me. The reader would be alarmed to know what, over several years, I put on the table in lieu of pasta. I had successes. I had failures. No one outside my immediate family was seriously harmed.

My happiest experiments were with corn-off-the-cob. Here is a reasonable example.

Corn-as-pasta: Stilton and Pecans
(For four, as a first course)

a little more than 1/2 cup pecans
a little more than 1/2 cup good, ripe Stilton cheese
4 cups of kernels cut from fresh sweet corn
2 Tb peanut oil
salt to taste

Toast the pecans,[1] then chop them roughly.
Chop the cheese.
Place on high heat a pan large enough to hold all the corn easily. When really hot, add the oil and after waiting for the oil to heat, throw in all the corn and stir to coat the kernels. Continue cooking until the corn is just done (3 minutes or less).
Now turn down the heat and add everything else except the nuts, stirring until the cheese melts. Salt to taste, sprinkle on the nuts, and serve.

[1] My method for toasting nuts: Place the nuts in a moderate oven. When they burn, remove and discard them. Place more nuts in the oven. Watch them carefully. Check every minute or so until they are fragrant and have changed color slightly. Remove the nuts.
Turn off the oven.

One could complain, "Corn isn't soft like pasta. And corn doesn't absorb flavors like pasta." Well, that's true. But this is a great dish.

If you want a soft faux pasta, rice works. I don't mean risotto, though. Risotto isn't faux pasta. The process of cooking real pasta is the same, mostly, no matter what sauce you use. A proper faux pasta should be something you just cook and then mix with sauce. With risotto, different recipes differ at the start.

The type of rice is important. Good jasmine rice can give you a soft but chewy grain that works well with many pasta sauces.

Rice-as-pasta: Butter and Sage

(For four, as a first course)

> 1 1/3 cups fresh jasmine rice
> 1 1/3 cups water (see below)
> 3 Tb butter
> 1/4 cup chopped fresh sage
> 1/3 cup freshly grated Parmesan cheese
> salt to taste but at least 1/2 tsp

I buy jasmine rice in 25-lb bags from a local Asian grocery store. About half the time the bags are labeled "new crop," or something like that. This is the best stuff (unless it's not really new). For this, you use one cup water for each cup of rice. If the rice is not so fresh, an additional tablespoon or two[2] of water per cup of rice seems to work.

Place the rice plus the right amount of water (see above) in a saucepan. Cover the pan and turn the heat up high. When the water starts to boil (but before it boils over) turn the heat down as low as possible, keeping the pot covered. The rice will bubble and steam. Turn off the heat when you no longer see steam when you look under the cover (but before the rice burns); this will be in about 5 or 6 minutes. Let the rice sit covered for another 5 minutes.

Place the butter in a large bowl. When the rice is ready, rake it out of the pan with a fork into the bowl. Toss the rice with the butter to distribute and coat the grains of rice. Now add the sage and cheese and mix. Salt to taste. I like to use sea salt, grinding the crystals if they are big.

---

[2] Or three or four.

I hope I haven't made cooking rice sound difficult. It's not difficult. You have a few minutes leeway in turning down the heat. And if the rice boils over, that's okay. It just makes a mess.

You also have a few minutes leeway in turning off the rice. And if the rice burns, most of it is still good. And I have a great recipe for the brown stuff at the bottom.

Other ideas? Chickpeas? Zucchini? Scallopini? There's no end to this.

There are more recipes[3] on the website:

press.princeton.edu/titles/10436.html.

And if *you* have ideas, I'm interested.

NOODLES AND DOODLES

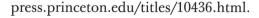

Apart from rhyming, noodles and doodles have nothing in common. I chose them to illustrate some shared features of mathematics and gastronomy, features that appear in this book again and again.

First of all, they are *pleasures*. Of course, sometimes we cook because we're hungry. And sometimes we calculate because we have to pay our taxes. But real cooks and real mathematicians play. They play with structures, they play with ingredients, they play with the ideas and the flavors that attract them strongly.

Second, while the attraction is aesthetic, it's also intellectual. We're curious. We want to taste; we want to tinker; we want to explore; we want to find out. We savor the unknown.

Third, and this may be the most important point, we often don't know what we're doing. We stumble around. Mathematics and gastronomy are mysteries. We have to stumble to make progress. We experiment. We try one thing. We try another. We may appear to have no method. But that's not true. Stumbling around *is* a method. It's the go-to method, surprisingly, of the best cooks and the best mathematicians.

---

[3] Including a recipe for burnt rice pudding.

Stumbling (and making progress) is the focus of the next chapter.

Hundreds of books are devoted to solving math problems. Thousands of books are devoted to cooking techniques. In the next chapter I will convince you (maybe) that the key to one is the key to the other.

2

THE ARROGANT CHEF

I SOMETIMES hear people say

"I can't bake bread."

The way they say it is familiar to me. The words are different, but the tone is the same.

"I can't do math."

The confessions are similar and similarly sad. They're not about weakness; they're about anxiety. In truth, everyone can do math and everyone can bake bread. Both acts are exercises in problem-solving. The remarkable fact is that the best method for solving math problems is also the best method for solving problems in the kitchen.

I have a simple theory about problem-solving. What you need to solve problems is a split personality. You need, first of all, *confidence*. Good problem-solvers are sure they can solve anything. Given a problem, successful problem-solvers dive in fearlessly, certain that they'll crack it right away.

But you also need *doubt*. Once you have a solution, the confidence has to step back. You need to question your answer, worry about it. Test it, tweak it. At the final stage, good problem-solvers act as though they're sure there's something wrong with their answers.

I think you can see the usefulness of both these personalities. You need the confident one at the outset. Without confidence, it's hard to begin. And if you do get started, lack of confidence can sap your strength and will.

And then, when you have an answer, you want the humble, doubting personality. You want to analyze your answer, understand it, check it for flaws.

The first personality is the most difficult to assume. You can't simply decide to be confident. I offer my students a substitute personality which seems to work pretty well. I tell my students to be *arrogant*. Most of them can do this.

BAKING BREAD

As an example, let's see how arrogance works in the kitchen. Let's talk about bread. So much has been written about bread that trying it for the first time can be scary. The difficulties are endless; there's so much to know, and so many ways to fail, or so it seems. How do you deal with yeast? How do you knead? How do you shape the loaves? How do you know when it's done? It takes years to learn!

Let's look at a simple recipe, and see how an arrogant (and ignorant) cook can bake bread successfully on the first try.

Very Plain Bread

1 package dry yeast
1/2 tsp sugar
1/3 cup lukewarm water
2 tsp salt
2 Tb oil
2 cups lukewarm water
5 to 6 cups flour
two 9-inch bread pans
butter for greasing the pans

Mix the yeast, sugar, and 1/3 cup water in a small container. Place the salt, oil, and 2 cups water in a large bowl. When the yeast mixture is foamy, add it to the bowl. Add flour, a cup or less at a time, mixing it in, until you have a stiff dough. Knead, then cover the dough with a damp cloth and set to rise. When it has doubled in bulk, punch it down. When it has risen again, divide the dough in two, shape into loaves, and place in the greased bread pans. When the loaves have risen about 50 percent, place them in a preheated oven at 425° and bake for 45 minutes. The bread is done if it sounds hollow when tapped on the bottom and if it slides easily out of the pan. Remove the loaves from the pans and cool on a rack.

That's all. It's a minimalist recipe.

The meek cook says: "It doesn't tell me how much flour to add or how to knead. It expects me to know these things already. I don't know these things. It's hopeless."

But the arrogant cook thinks: "If the recipe doesn't tell me things, it must mean they aren't very important. I'm going ahead to see what happens."

Let's be arrogant and try it.

We have no problems until we add the flour. We don't have bread flour. We're using all-purpose flour (this is a purpose, isn't it?). We add flour and after a while it gets hard to stir. Maybe it's time to knead.

We have a hazy idea of kneading. Maybe we've seen an "iron chef" do it, or maybe we saw something on YouTube. In any case, we put our hands into the bowl and start to mess around. It's pretty sticky. It doesn't seem right. Maybe it could use more flour. We put in more flour. We wrestle with it some more. Maybe we dump it out on the counter and maybe we don't. After a while, we decide that's all the kneading it's going to get. The dough looks untidy. We set it aside to rise.

Rise where? In the bowl? On the counter? Under the bed? The bowl looks bad. There are bits of flour sticking to it, so we leave the dough on the counter.

Somehow, we stumble through the recipe. It doesn't exactly rise, it sort of spreads out. It takes hours. We keep going. We make crude loaves. We bake them. How did they turn out?

Actually . . . the bread came out pretty good.

Why? In all honesty, we did a lousy job. We didn't knead the dough enough. Then the dough dried out when we left it on the counter. The recipe we used was basic, almost primitive. We put nothing exciting or flavorful into the dough. *Why did it work?*

Well first of all, supermarket bread is *terrible.*[1] We have to fail very badly indeed to produce something worse. No matter what we do, our bread has the incomparable taste and aroma of yeast. It's real.

[1] It's better than it used to be. But what has improved most dramatically is the pretention. What was "premium" is now "artisanal."

Second, bread is extremely forgiving. You can use more yeast or less yeast, more sugar or less sugar, more oil or less oil. You can knead it longer or knead it less, let it rise longer or less, bake it longer or bake it less. You'll still make pretty good stuff.

Strange things can happen, of course. I've made bread that flowed out and over the rim of the bread pan. Prying it loose from the pan was a battle. It looked ridiculous. But it tasted great.

And what if we had failed?

In the event of failure, we would have wasted ingredients worth at most a dollar. We didn't spend much time. We didn't disappoint anyone (we weren't so arrogant that we invited guests). And we learned something.

Every time we cook we learn. Wearing our humble personality, we reflect on what we liked and what we didn't like. We look for more information, recipes, and descriptions. We plan what we might do to improve the bread.

I bake bread several times a week. I've been doing it for 25 years. My basic recipe took ages to develop. It's fabulous and I say so myself.

I'm going to give you the recipe here and will humbly accept your praise. But part of me hopes that you will ignore it. That part of me wants you to spend years in your own quest for bready perfection. If you do, you'll have fun (and eat well). And you'll end up with a loaf that is new and wonderful.

A little explanation before I give you the recipe. I wanted my bread wheaty but light. I wanted it healthy but entertaining. I wanted it flavorful but undemanding. It's mostly white flour, so it has a light texture. The wheat germ gives it the protein of whole wheat bread. The other items provide additional wheatiness.

The recipe is designed for taste and convenience. It takes me 15 minutes of work altogether, time that can be scattered throughout the day at my convenience. It's no trouble to supply a household with fresh bread every day.

My Bread

    1 tsp dry yeast
    1/2 tsp sugar
    1/3 cup lukewarm water
    2 tsp salt
    2 Tb oil
    2 cups lukewarm water
    1/2 cup whole wheat flour
    1/4 cup toasted wheat germ
    1 tsp sesame seeds
    2 to 4 Tb cheap balsamic vinegar
    2 to 4 Tb crumbs (optional, see note)
    flour
    two 9-inch bread pans
    butter for greasing the pans

The method of mixing, rising, and baking is the same as for the earlier recipe. You add the extra ingredients with the flour.

Notes:

- I don't give a quantity for the flour. It will be around 5 or 6 cups, but (see below) you have a lot of leeway. Start with less and add just enough so it is dry enough to knead. Add more, if you like, when kneading.

- I mix, knead, and set the dough to rise in one very large bowl. I don't clean the bowl between kneading and rising. Why? I'm lazy (see Vanity and Sloth in chapter 8). Also, it doesn't seem to matter.

- The first purpose of kneading is to mix the dough. You should be able to tell easily when you've accomplished that. The second purpose is to develop the gluten, which increases the stretchiness of the dough. This can be important for some uses, but for loaf bread, a minimal amount of kneading gives you all the stretchiness you need. Long rising also helps develop the gluten.

- It's amazing what effect the sesame seeds have. Only the ones that reach the outside of the dough get toasted, but these add a lot of flavor.

- The vinegar was suggested by a British colleague, Richard Kaye. He uses malt vinegar (tasty and cheap where he lives) to achieve a slight

continued

continued

sourness. Malt vinegar is expensive where I live. The balsamic vinegar I use is pretty cheap. I don't think I use enough for detectable sourness, but it adds some flavor and also helps the yeast do its work.

- Almost all the parameters of the recipe are negotiable. You can knead 60 seconds or 10 minutes. The dough can rise in a cold house or a hot one (cold takes longer, that's all). You can let the dough rise once, twice, three, or even four times. You can bake it 40 minutes or an hour. Basically, *you can't mess this up*.
- Well, maybe you can. If the dough rises much too long (extreme neglect), it can get sour. This is less than wonderful.
- Also, if the dough rises too long in the pan, it can collapse and your bread will be sunken-looking and tough.
- Altogether, this bread has the nutrition of dense whole wheat bread, except for the fiber. To get more fiber, eat more bread. That's what I do. Specifically:

Nutrition Facts

Serving Size: 2 slices
Servings per loaf: about 8

Serving Size: 2 slices	
Calories 162 Calories from Fat 7	
	% Daily Value
Total Fat 8 g	1%
Saturated Fat 0g	
Polyunsaturated Fat 0g	
Monounsaturated Fat 0g	
Cholesterol 10mg	
Sodium 236mg	1%
Total Carbohydrate 32.5g	11%
Dietary Fiber 1g	4%
Sugars less than 1g	
Protein 4.5g	

Vitamin A 0%	Vitamin C 0%
Calcium 0%	Iron 10%
Thiamin 30%	Riboflavin 10%
Niacin 6%	Vitamin B_6 8%
Folic Acid 25%	Vitamin B_{12} 30%
Phosphorus 30%	Magnesium 20%
Zinc 30%	Copper 4%

- I have to explain the crumbs. The bread comes out of the oven with a beautifully dark crust. When you slice it, a lot of crumbs are produced, crumbs that contain the most flavorful molecules of the bread. I hate to see these go to waste, so I collect them and use them from time to time in other recipes. One day, I realized they could enrich the bread itself.

- Part of the savoriness of the crumbs comes from the sesame seeds. When adding crumbs, I use only a half teaspoon of sesame seeds.
- Adding flavor is reason enough to use the crumbs, but as an additional benefit, I can add to the nutrition statement the alarming notice:

This product contains 4% (post-consumer) recycled materials.

- I often hear people say that they tried to make bread once and the result was awful. Of course, they didn't try *my* recipe. Even so, I'm still prepared to dispute them! Perhaps the bread didn't come out as well as they'd hoped. So what? They made bread. And for a while, I'm pretty sure, it was fresh bread. How can you beat that?

And when you bake bread, you should love your bread. It's *your* bread!

Say your kid brother comes home from school with a C in mathematics. You still love him, don't you?

The Nobel Prize–winning physicist Richard Feynman was a paragon of confidence. He was intellectually curious but curiosity is not uncommon. He was brilliant but many people are brilliant. What set Feynman apart was his fearlessness. He felt he could tackle any problem. To him, ignorance was no obstacle. His wonderful memoir *Surely You Must Be Joking, Mr. Feynman*, contains many examples of this—cognition, myrmecology (ants), sexology, locksmithing, urology.

That's the difference between a great scientist and an ordinary genius:

Confidence.

SOLVING PUZZLES

Let's see how an arrogant attitude helps the mathematical problem-solver. We'll tackle a puzzle I've chosen because it illustrates the usefulness of having a chip on one's shoulder.

The puzzle is based on "Spin-Out," a mechanical puzzle sold in stores in several versions (one has elephants). Spin-Out consists of a sliding plastic strip with knobs inside a case.

Since you probably don't have Spin-Out with you, we'll look at a different puzzle, one that is structurally the same. We'll call it "Flip-Out."

It starts out with seven coins in a row, all heads-up.

The object is to get all the coins tails-up. The rule is that you can either flip

- the farthest coin on the right, or
- the left neighbor of the rightmost head

For example, in the position

you may flip either coin 7 (the farthest coin on the right) or coin 5 (because coin 6 is the rightmost head).

Let's watch the arrogant problem-solver (we'll call her Smedley) tackle Flip-Out.

At the start, there are two possibilities, flipping coin 7 or coin 6. What does Smedley do? After a moment's hesitation, Smedley flips coin 7.

Why?

"No reason, really," says Smedley, "I had to do something." What next? Hmm . . . Smedley can flip either 5 or 7.

Flipping 7 returns the puzzle back to the start. That's not for the arrogant problem-solver! The arrogant problem-solver never turns back![2]

---

[2] Anyhow, not this soon.

So Smedley flips 5.

Does Smedley have any idea where this is leading? "No idea!" says Smedley.

Smedley can now flip either 5 or 7. Flipping 5 undoes the last flip, so Smedley flips 7.

Smedley just keeps chugging along, not thinking very hard. Maybe you can see what's happening. There's a clue in the rules for the puzzle. The rules say that at every stage you can flip either

- the farthest coin on the right, or
- the left neighbor of the rightmost head

That's just two possibilities. If you think about it a little, you'll see that no matter what you flip, you can always unflip it on the next move. In other words, there are always just two directions. Smedley chose one direction, flipping 7, then 5, then 7. In the other direction, she could have flipped 6 then 7, then 4. At every stage you can keep going or you can turn around.

The Flip-Out puzzle is really like the following maze,

except it's disguised. Once you get started, you never really have a choice; all you can do is keep going or turn back.

If you've been playing with some pennies, you'll know what happens to the arrogant problem-solver here. If you follow one way you get to the Finish, All Tails.

If you follow the other way you get to a dead end, in this case:

Smedley will solve this puzzle, either on the first try or on the second. She'll solve it because she doesn't turn back.

Many people fail to solve this puzzle.

Why? They go in one direction for a while, then they begin to worry. "Am I making a mistake? I don't seem to be getting anywhere! Maybe I should go back!" They shift directions. Unfortunately, they're no more confident than before. After a while, they change directions again and so on, back and forth, back and forth. Eventually, they give up.

The problem is a lack of arrogance!

But maybe I should apologize for all the arrogance in this chapter. *You're* not arrogant, I know that. You're amiable, modest, considerate. You would never think of behaving arrogantly. Smedley must be hard for you to take.

But if you can manage a sort of mock-arrogance you'll find it helpful. Attack problems (and recipes) as you imagine Smedley would, confident even when clueless.

And we move, in the next chapter, to aesthetics.

What, exactly, is the beauty of mathematics? It's an elusive concept, subtle, abstract and intellectual. I've struggled for years to describe that beauty. But in time, the answer came to me.

In a word, it's *cheesecake*.

3

SIMPLE TASTES

ALL OF us, at times, crave food with simple, uncomplicated tastes—pancakes, for example, a boiled egg, hamburger, banana, corn on the cob, or root beer on the rocks. Sometimes we want familiar, natural, honest fare.

Simplicity is one of many qualities that can be attractive in food. Simple ingredients, simply prepared. Food without artifice. Food that is what it appears to be.

Simple food is not necessarily unsophisticated. There are at least four cookbooks in my house with "simple" in the title.[1] Simplicity can be difficult to achieve. Consider, for example, a dish of homemade fettucine with butter and Parmesan. It is utterly simple. I'm still working on it.

Simplicity is a mathematical aesthetic too. Mathematics, simply constructed and simply presented, can be attractive and compelling. A simple geometrical figure with a puzzle, a game with simple rules, a mathematical world with just a few simple assumptions—these can pull you in. They're like a door opened wide through which one can see a lovely and mysterious garden.

Complexity is attractive too. That's in the next chapter.

A SIMPLE FOOD

Cheesecake embodies the beauty of simple food. The best cheesecake (my opinion) is unadorned, free of extraneous elements. There is just the clean, sweet/sour taste of milk and cream cheese. And maybe a crust.

I grew up on one kind of cheesecake, the thin sort, baked in a pie dish and topped with a custard of sweet/sour cream. Probably I ate

---

[1] Two of the "simple" cookbooks are by Jean-Georges Vongerichten and neither of them is simple.

a lot of Sara Lee cheesecakes. They were good. But when, as a young man, I encountered New York–style cheesecake, I was blown away. I was eating the essence of cream.

I was in Boston, though. I met this class of cheesecake in the Baby Watson version, which I think is creamier and not as sour, not as dry, as true New York cheesecake. I really liked it and my wife liked it. But Baby Watson was expensive.[2] We decided to figure out how to do it ourselves.

There are innumerable recipes for cheesecake. Many of them are described as the "best." The following recipe, though, is actually the best.

Cheesecake

A large springform pan, 3 inches deep, is needed for this cake. Also very useful is a large bowl mixer. It's important to cook this cake a day in advance (it's handy, too).

For the crust:

    1 1/2 cups flour
    1/2 tsp salt
    1/3 cup sugar
    1 stick unsalted butter

Sift together (or put through a strainer) the flour, sugar, and salt. Cut in the butter with a pastry cutter until the butter is in small bits, pea-size or smaller. A pastry cutter is an efficient gadget that slices the fat up. Lacking one of these you can always use a knife or a pair of knives, but it takes longer. A food processor is efficient for this, but you have to stop it before the fat is completely chewed. You want pieces of fat about the size of peas. Press the mixture into the bottom of the pan. Bake at 350° until aromatic and slightly golden (15 minutes or more). Check frequently.

For the filling:

    2 pounds (4 bricks) cream cheese
    1 1/2 cups sugar
    4 eggs

---

[2] I'm cheap. See chapter 8.

1 1/2 cups heavy cream
1 tsp vanilla extract

Cream the cheese and sugar. Beat in the eggs, one at a time. This is where the bowl mixer comes in handy. Beat in the cream and vanilla. Pour into the pan containing the baked crust. Turn the oven down to 300° and bake until the filling rises and starts to brown around the edges. This will take an hour or so. Check frequently. When it's done it will still wiggle when jiggled.

Turn off the oven, but leave the cake inside. Open the oven door slightly and allow the cake to cool gradually.

After 20 minutes or so, you can take the cake out of the oven to cool further. When much cooler, cover with plastic and refrigerate. The next day, when you remove the plastic, you'll probably find some moisture, which you can remove with a paper towel.

Now let's talk about the practice of pouring a sweet, fruity sauce over a slice of cheesecake. This makes no sense here. The quiet flavors of cheesecake are totally obscured by goo.

A sweet sauce might make sense for a cheesecake with distinct sourness. But my mathematical soul loves the simplicity and purity of (my) cheesecake filling, a filling that whispers "cream, cream, cream." The purity is enhanced by the crust, which, unlike the noisy graham cracker version, says quietly, "butter, butter, butter." That delicate duet would be drowned out by a sauce.

But if you want to tinker with flavor, the following is extremely good.

Almond Cheesecake

The recipe is the same as the basic recipe except the crust is made with

3/4 cup ground almonds
3/4 cup flour
1/2 tsp salt
1/4 cup sugar
3/4 stick unsalted butter

continued

A SIMPLE PUZZLE

I want to tell you about some research I did with a student, Colleen McGaughey, with a friend, Jerry Butters, and with my son, Fred Henle. The work was motivated by a desire for mathematical simplicity.

A few years ago I was thinking about sudoku puzzles and it seemed to me that there was something untidy about the numbers scattered about the puzzle.

		1		2	4			9
		3		9		8		5
			3			4	2	
4			2					
5				6				7
					9			8
	6	4			2			
9		2		7		5		
7			6	4		9		

It's not beautiful. And the numerical values play no role in the puzzle. If we substitute *a* for every 1, *b* for every 2, and so on,

		a		b	d			i
		c		i		h		e
			c			d	b	
d			b					
e				f				g
					i			h
	f	d			b			
i		b		g		e		
g			f	d		i		

we get an equivalent puzzle. I felt that

1. Numbers in a puzzle should be meaningful. Maybe somewhere they should be added, or multiplied, or something. If you're going to have numbers, you should have arithmetic.
2. The numbers should be invisible at the start. That is, there shouldn't be number clues. The puzzle would be nicer, cleaner, if the numbers in the answer could come from clues in the puzzle that were numberless.

Fred, Jerry, Colleen, and I thought about this and we came up with the idea of drawing odd-shaped regions in the square. Then we would require that all the regions have the same sum.

It took us a while to find a puzzle of this sort. Like all good puzzles, our puzzle had to have a unique solution. Everything we tried either had no solutions or many. Finally we found this, a 6 × 6 puzzle:

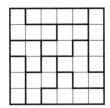

No numbers! Isn't it beautiful?

The rule for rows and columns is the same as for sudoku—you can't repeat numbers. For a 6 × 6 puzzle, you just use the digits 1, 2, 3, 4, 5, and 6, so in the answer, each row and column should have exactly one of each digit. And the sum of the numbers in each region must be the same.

How do you solve a puzzle like this? How do you get numbers?

The key is that you can compute the sum of all the numbers in the square. The sum of the numbers in any row is

$$1 + 2 + 3 + 4 + 5 + 6 = 21.$$

That means the total sum of the numbers in the entire square (six rows) is

$$21 \times 6 = 126.$$

Since there are nine regions, all with the same sum, the sum of each region must be

$$126 \div 9 = 14.$$

Now look at the straight, three-cell region at the bottom left. The combinations of three numbers adding to 14 are

    6 + 6 + 2   6 + 5 + 3   6 + 4 + 4   5 + 5 + 4.

Only 6, 5, 3 works here, though, because you can't repeat numbers in a column.

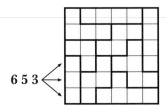

What about the three-cell region just above? The square on the left,

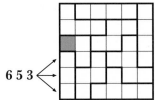

can't be 6, 5, or 3. It can't be 2 or the other squares in the region would both be 6's. A little more thought shows you the only way to fill the region is with two 4's and a 6.

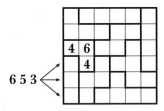

Thus, the numbers at the top of the left column must be 1 and 2.

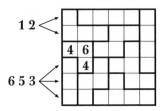

That tells us that the remaining numbers in the top left region must be 6 and 5.

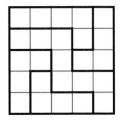

Notice that as we solve the puzzle, we're producing a (mathematical) proof that the solution must be what we say it is.

I leave the rest to the reader. It gets harder, but it stays fun.

I have to say we became a little obsessed with these puzzles. We called them "clueless sudoku" puzzles. Of course they're not clueless; it's just that they have no numerical clues.

Some have suggested that these puzzles are really "clueless Ken-Ken" puzzles. Perhaps that's more accurate. KenKen puzzles are just as inelegant as sudoku; actually, they're worse. Not only do they have numbers, they have operations. Scandalous.

Clueless sudoku puzzles are hard to find. If you play around a bit you'll see that there are no 2 × 2 puzzles. There are no 3 × 3 puzzles, either.

There is a 1 × 1 puzzle.

But it's not very interesting. We found a 5 × 5 puzzle.

We stewed for months over the question of a 4 × 4 puzzle. Each row of a 4 × 4 adds to 10.

$$1 + 2 + 3 + 4 = 10.$$

So the total of all the numbers in the square is 40.

$$4 × 10 = 40.$$

That gives us a lot of possibilities. We could have

1 region	with a sum of	40
2 regions,	each adding up to	20
4 regions,	each adding up to	10
5 regions,	each adding up to	8
8 regions,	each adding up to	5
10 regions,	each adding up to	4
20 regions,	each adding up to	2
40 regions,	each adding up to	1

Some of these are ridiculous. Clearly we can't have all regions adding up to 1 or 2 since some of the numbers in the square are 3 and 4. And if we have just one region, then there certainly won't be a unique answer.

We can pare the possibilities down to these:

2 regions,	each adding up to	20
4 regions,	each adding up to	10
5 regions,	each adding up to	8
8 regions,	each adding up to	5
10 regions,	each adding up to	4

The last one is impossible and here's why. If every region adds up to 4, then what do you do with a 2? The only way is to combine it with two 1's, $2 + 1 + 1 = 4$. But there aren't enough 1's to handle all the 2's.

You also can't have all the regions add up to 5. This is a little trickier. The reason is that every 4 has to be matched with a 1.

$$4 + 1 = 5.$$

That's the only way to get a sum of 5 using a 4. A solved puzzle might look like this.

2	1	4	3
3	4	1	2
4	2	3	1
1	3	2	4

But then you could switch the 4's and the 1's to get another solution.

2	4	1	3
3	1	4	2
1	2	3	4
4	3	2	1

Puzzles must have unique answers. That's part of the aesthetic of simplicity. So now we're down to just three possibilities.

2 regions,   each adding to   20
4 regions,   each adding to   10
5 regions,   each adding to   8

The first of these doesn't look promising. The other two look reasonable.

We spent a long time searching for a four-region and a five-region puzzle. In the end, a computer program we wrote proved that there is no 4 × 4 clueless sudoku with four or five regions.

Imagine our surprise, then, when we found a puzzle with two regions!

There's only one solution![3] Beautiful!

The math and the cooking in this chapter weren't simple as in "easy." They were simple as in "not ornate." Simplicity is good.

Complexity is good too . . .

[3] See press.princeton.edu/titles/10436.html for the solution to this and the other puzzles in this chapter.

4

COMPLEX FLAVORS

COMPLEX FOOD

Grape juice is good. So is wine.

Grape juice is a simple drink. You know the taste. Every sip is the same, 24/7. That can be just what you want if you're thirsty, if you need energy, if you only have a five-minute break.

In contrast, wines have a multitude of flavors. Even in a single bottle, the taste can change as you drink it, as the wine warms, as it alters chemically with exposure to air. For practiced tasters, a succession of sensations can be noticed in a single sip.

Complexity, of course, is not enough. The flavors of a dish must support and complement each other. In *coq au vin*, a masterpiece of French cuisine, the flavors of chicken, both concentrated (brown stock) and freshly cooked, combine with those of burgundy, tomato, garlic, bacon cooked in butter, sweet onions cooked in butter, mushrooms cooked in butter, and flamed cognac.

Coq au vin is not a simple dish but it's important to note that simplicity and complexity are not incompatible. A great dish can be both. Consider, for example, fruit. A single apple has to be considered a simple dish. But as anyone who has experienced that fruit just plucked from a tree knows, apples can offer a complex array of tastes.[1]

Fruit is a nice example of simplicity/complexity. Factory-farmed strawberries are not complex. They're bred for shelf life and consistency. The flavor is stable and one-dimensional. Fresh local strawberries, on the other hand, offer a variety of tastes, levels of sweetness, and tartness.

[1] My favorite: Ashmead's Kernel, an eighteenth-century apple, first grown in Gloucester, England.

But with skill, complexity can be achieved even with mass-produced strawberries. Marcella Hazan[2] was an apostle of simplicity. But simplicity and complexity can go hand in hand. She suggested combining the berries with balsamic vinegar and sugar.[3] That works beautifully.

I once served apparently plain strawberries at a dinner in March. My guests were pleased and asked me how I got such fresh strawberries in March. What I had done was sprinkle the berries with a little Grand Marnier. For some reason, the orange flavor of the liqueur added just enough complexity to fool guests into thinking they were eating something special. I hadn't planned to deceive. I just thought the berries would taste better.

The complexity of raspberries is brilliantly brought out in a pie recipe I was given at a pick-your-own raspberry farm. The secret is a combination of cooked and fresh raspberries. I have fiddled with the recipe only slightly.

Quonquont Raspberry Tarts

(6–8 portions)

For the tart shells:

1 cup flour
2 1/2 Tb sugar
1/4 tsp salt
2/3 stick unsalted butter

Sift together the flour, salt, and 2 1/2 Tb sugar (or use a wire whisk to combine them). Cut in the butter with a pastry cutter until the butter is in fairly small bits. The dough will be dry and crumbly. Divide it among individual tart pans and press it down evenly. Bake these at 350° until they are fragrant and golden. When cool, remove the crusts from the tart pans and place on dessert plates.

For the filling:

1 quart raspberries
3/4 cup water

continued

[2] Marcella Hazan (1924–2013) was a teacher and prolific cookbook author. She was responsible, more than anyone else, for opening up Italian cuisine to Americans.

[3] *Marcella Says . . .* (HarperCollins, 2004).

continued

7/8 cup sugar

3 Tb cornstarch

Put a cup of the berries in a saucepan with the water, bring to a boil, and reduce heat to a simmer. Break up the berries with a wire whisk. Mix the sugar and cornstarch thoroughly, then add to the cooked berries, while stirring vigorously with a whisk (you want to avoid forming lumps of cornstarch). Cook the mixture until it changes from watery and cloudy to thick and clear (but dark). Remove from heat, and cool to just warm.

When the mixture has cooled but not completely jelled, add the uncooked berries and mix. Place the mixture in heaps on the rounds of crust.

For the topping:

1 cup cream for whipping

1 Tb sugar

1/4 tsp vanilla

Whip the cream, add the sugar and the vanilla and whip a little more. Top the tarts with the cream and serve.

Notes:

- Ideally, this dessert is put together not more than an hour or so before serving.
- If you don't have individual tart pans, you can cook the crust in a large tart pan or pie pan. Divide the finished crust into (necessarily irregular) pieces for the dessert plates.
- The original Quonquont farm recipe was for a raspberry pie. The mix makes a sensational pie, but you need to assemble it just before serving or the filling will make your crust soggy. The crust given here resists sogginess for a while, but not forever.
- For additional complexity you can add to the cooked berries a teaspoon of crème de cassis.

COMPLEX MATHEMATICS

I'm going to look at a few games. Games are a recognized area of mathematics. But they are more than that. They are a perfect microcosm of mathematics. They can be theoretical and they can be applied. They can be analyzed and they can be played for enjoyment. They can be simple and they can be complex.

A particularly simple game is what I will call "1-2-3 Takeaway." You start with a pile of sticks.

Two players take turns removing sticks. They can remove 1, 2, or 3 sticks. The last player to move wins the game.

Clearly if there are only 1, 2, or 3 sticks in the pile,

the first player (let's call her player I) has a winning strategy, namely, take all the sticks.

If there are exactly 4 sticks in the pile,

the second player (call him player II) has a winning strategy—because no matter how many sticks player I takes, player II can take the rest and win.

This is a simple game because it's easy to analyze the game for any size pile. Suppose, for example, there are 23 sticks.

Arrange them in bunches of 4, with 3 left over.

Then player I should take the extra 3. That leaves five bunches of 4. From now on, player I should

- take 1 stick if player II takes 3 sticks,
- take 2 sticks if player II takes 2 sticks, and
- take 3 sticks if player II takes 1 stick.

That way every pair of moves eats up one of the bunches of 4. Player I will win.

This analysis shows that player II has a winning strategy if and only if the number of sticks is divisible by 4.

number of sticks in the pile	1	2	3	4	5	6	7	8	9	10	11	12	13	14	...
player with the winning strategy	I	I	I	II	I	I	I	II	I	I	I	II	I	I	...

The pattern is easier to see if we represent games where player I has a winning strategy with a colored square and games where player II has a winning strategy with a white square.

Now suppose we change the game to allow players to take 1, 2, 3, or 4 sticks. Again, the last player to move wins.

This game is just as easy to analyze. Player I wins if the pile has 1, 2, 3, or 4 sticks. Player II wins if the pile has 5 sticks. As before we have a repeating pattern. this time it is:

number of sticks in the pile	1	2	3	4	5	6	7	8	9	10	11	12	13	14	...
player with the winning strategy	I	I	I	I	II	I	I	I	I	II	I	I	I	I	...

or

Now a more interesting game results if we limit players to 1, 3, or 4 sticks. As always, the last player to move wins.

Player I wins the 1-stick game. Player II wins the 2-stick game. Player I can win the 3- and 4-stick games. What about the 5-stick game?

Player I can win this too by taking 3 sticks (leaving 2 sticks, which is bad for player II). And player I can win the 6-stick game by taking 4 sticks. But player II wins the 7-stick game. Again we have a pattern, but a more interesting one. It repeats every seven squares.

1 2 3 4 5 6 7...

These games are sometimes called "Nim" games. Individually they're not complex. No matter what moves are allowed, you always get a repeating pattern like the ones above. But surprisingly, no one has yet found a way to find that pattern, except by laboriously checking piles and strategies, one by one. There are mysteries in Nim.

Now here's my especially complex game. I call it "Nimrod."[4]

As before, we start with a pile of sticks. Player I can take 1 or 2 sticks. From then on a player has (usually) three choices:

- she can take as many sticks as her opponent did,
- she can take one stick more than her opponent did, or
- she can take one stick fewer than her opponent did.

I said "usually" because you always have to take at least one stick. As in the other Nim games, the winner is the player to make the last move.

There is a theorem that for any finite, two-player game, one of the players has a winning strategy.[5] That means that we can produce a sequence of squares for this game. Here is how it starts:

1 2 3 4 5 6 7...

---

[4] My son Fred's suggestion. You'll see why at the end. It does and doesn't have anything to do with the biblical king of Shinar.
[5] See the website for a proof of this. It's not hard.

This game is seriously complex. To convince you of that, I'll show you a more detailed picture. At any point in the game there are two important numbers: the size of the pile and the current rate of stick removal. The picture below shows who has the winning strategy for different pile sizes and removal rates. As before, a red square means player I has a winning strategy and a white square means player II has a winning strategy.

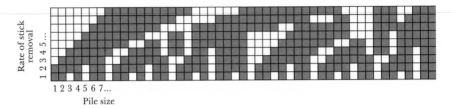

For example, the square for pile size 5, removal rate 2 is white. That means that player II has a winning strategy in that situation. It's not difficult to see what that is. If I takes 2 sticks, II can take 3 sticks and win. If I takes 3 sticks, II can take 2 sticks and win. And if I take 1 stick, II can take 1 stick, leaving player I again with a white square.

The picture above is messy. We can see some patterns if we look at more data.

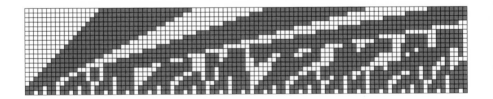

And here's a picture with lots of data.

Complex!

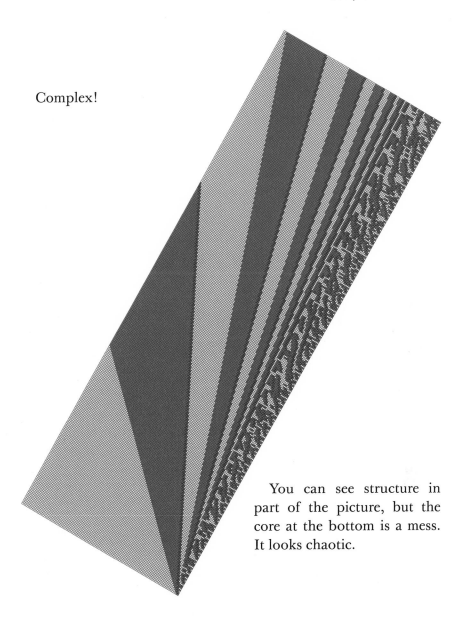

You can see structure in part of the picture, but the core at the bottom is a mess. It looks chaotic.

I have to admit that I don't understand Nimrod! And to the best of my knowledge, no one understands it.

Before I end this chapter, I want to tell you about three variations that are a lot of fun. The first could be called "Chicken."

Instead of a pile of sticks, imagine two hot rods facing each other at the ends of a street.

The street is measured in car lengths. In Chicken, the drivers (the players) take turns moving forward. The first move for either player must be to go forward one or two car lengths. Thereafter, each driver has three choices:

- maintain speed (drive as far as he or she did before),
- accelerate one car length, or
- decelerate one car length

—except, as in Nimrod, drivers must move at least one car length forward. Crashing is not allowed and the winner is the driver to make the last legal move.

You can see how this is related to Nimrod (and you can see why Fred gave it that name). The distance between the cars is like the number of sticks in the pile. The games are different, though. Instead a single rate of stick removal, each player has his or her own rate.

The second variation I call "Jousting." Two knights on chargers face each other.

It's just like Chicken except that the object is to land your knight on top of your opponent's knight. You might, in the course of the game, run past your opponent (the game is played on an infinite board). In that case, both knights will slow down, stop, reverse direction, and charge again. In Jousting, you are allowed to stay motionless one move, indeed, you can't reverse direction without doing that. If one knight decides to run away, that player loses.

And finally, one more variation. This one is "Demolition Derby." It's a two-dimensional game. There's a detailed description on the website. It can be played by two or more players, each driving a car. Despite the cars, it's really a two-dimensional version of Jousting, that is, the object is to land on another vehicle. If you do that, you disable it; it's out of the game. The object is to have the last car that still moves.

I've given you simple and I've given you complex. Maybe you preferred simple and the game Nim. Maybe you preferred complex and the game Jousting. Maybe you didn't like any of them.

Whatever your preference, you have taste. That's what the next chapter is about.

5

THE DISCRIMINATING EATER

AT RESTAURANTS, you're in charge. You decide what you want and how you want it. You have taste, and your preferences matter.

You probably think that mathematics is different. But it's not different (except that there aren't restaurants). Specifically, I claim:

1. There are different flavors of mathematics.
2. Everyone (you included) has mathematical taste.
3. You (and everyone else) can choose your mathematics.
4. You don't have to feel guilty or inadequate about it.

Let's take these one at a time.

1. THERE ARE DIFFERENT FLAVORS OF MATHEMATICS.

Here's one flavor: **Algebra**. The particular algebra you studied in high school is just one example. There are many others.

By "algebra" I mean a system of symbols and rules. In high school you have infinitely many symbols (all those numbers, plus letters, plus +, etc.) and a lot of rules. You can have the same set of symbols but with fewer rules, with more rules, with different rules.

Here's a different rule:

$$\text{For all } x, y, \text{ and } z, \ x(yz) = (xy)(xz).$$

Surprisingly, that rule with just one symbol produces a pretty interesting algebra. Let's say the symbol is a and we'll need parentheses too, I guess. Then we have

$$aa$$
$$a(aa) = (aa)(aa)$$
$$(aa)(aa) = ((aa)a)((aa)a)$$

and so on. This algebra prompted some mathematicians to wonder, is there any expression *xxxx* which is equal to itself times something else? That almost happens here—

$$a(aa) = ((aa)a)((aa)a)$$

—but not quite. The answer turned out to be no, but it took years of work.

Here's another flavor: **Geometry**. Euclidean geometry, the geometry you studied in high school, is just one example. There are many others.

By "geometry" I mean a system of diagrams and pictures with meaning attached.

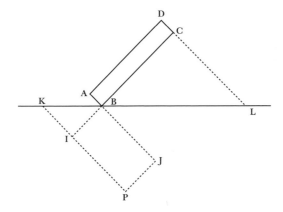

I use the word loosely. Many areas of mathematics have pictures that go with the ideas.

Here's another flavor: **Finiteness**.

There are finite algebras and there are infinite algebras. They're different.

There are finite geometries and infinite geometries. They're also different.

Interestingly, infinite geometries and infinite algebras were discovered well before finite geometries and finite algebras. Infinity sounds more complicated. It's easier to define "infinite" than "finite." You don't believe me? Try defining "finite."

There are many more flavors—numbers, shapes, motion, games. Some mathematical fields seem like nothing more than stripped-down logic. Some will appear later in this book.

2. EVERYONE (YOU INCLUDED) HAS MATHEMATICAL TASTE.

I haven't met anyone who didn't have an answer to the question, Which do you prefer, algebra or geometry? Symbols or pictures?

That's taste. Your choice might reflect a particularly good teacher or a particularly bad one. But almost certainly part of your preference rests on personal taste, or to use another word, aesthetics. One of those two subjects just seems *nicer* to you.

How do you feel about infinity? Some people are drawn to it. Some great mathematicians were repelled and frightened by it.[1] Some people find finite structures tidier, simpler. Some people find infinity cleaner, more beautiful.

Of course, personal taste is complicated. Geometry is more appealing to me than algebra, but most of my work has been in areas where pictures are really not helpful. And I'm strongly attracted to infinity, but while I've spent years studying infinities, my most recent work is finite.

3. YOU (AND EVERYONE ELSE) CAN CHOOSE YOUR MATHEMATICS.

This must seem crazy to you. "How can there be choice? Don't you have to use the mathematics that *works*? Don't you have to choose the math that's appropriate to the problem?"

But what if there is no problem?

I'm interested in mathematics for its own sake, not for what it can do. I take pleasure in mathematics apart from any application. It's the same in cooking. Restaurant chefs and diners are interested in dishes for their own sake, not (usually) for their nutritional value.

And what if there *is* a problem? (Or what if the nutritional content of your dinner is important?) You still have a choice, but if you want to solve the problem, you will, I'm sure, choose the appropriate mathematics or the appropriate dish. In that sense, by definition, statement 3 is still true!

But I want the reader to focus on the intrinsic satisfaction (or lack of it) of a piece of mathematics. You like it or you don't, just like a salad, a casserole, or the soup of the day. And then it's your choice—to eat it or not.

Do you like doing sudoku? That's a choice. Everyone can choose.

[1] Blaise Pascal is a famous example.

4. YOU DON'T HAVE TO FEEL GUILTY OR INADEQUATE ABOUT IT.

You don't like anchovies? Are you embarrassed by your relationship to anchovies? Do you think that your dislike of anchovies reflects badly on you? I'll bet you don't. And are you defensive about preferring Italian to French, Thai to Indian, or salty caramel latte to pumpkin pie green tea latte? Of course not.

The same should be true for mathematics!

Often people will tell me (hoping, I think, to end the conversation) "Oh, I was terrible at algebra!" In truth, they weren't really terrible. They just didn't like it. If they had liked it, they would have mastered it.

Anyone can learn algebra. But not everyone does learn algebra. If you don't like a subject, you won't take the energy to succeed. If you fear a subject, you won't have the courage to persist.

I won't argue with anyone that they should love a piece of mathematics. I will argue that they shouldn't fear it.

Let me make one more point.

5. PROFESSIONALS CHOOSE TOO.

Professional mathematicians have taste. They have preferences and passions. Most of them ignore vast areas of mathematics and just work on the stuff that floats their boats. They're having fun. They make no apologies and they have no regrets.

IN CONCLUSION

I'm urging you to develop your own taste. There's math in this book, a little in almost every chapter. Taste the bits as they come by. If a bit doesn't appeal to you, let it go. Skip over it. (Same with the food.)

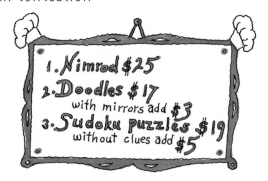

But if it tickles or intrigues you, read on.[2]

The next chapter is all about how I was intrigued and then later, tickled.

[2] And read more at press.princeton.edu/titles/10436.html.

6

THE PERSISTENT COOK

IF YOU want something and you believe you can get it, you'll strive for it. If you strive for it and it's achievable, you'll achieve it. That's true in every field, in particular, in cooking and in math.

A GASTRONOMIC CHALLENGE

When my son Fred turned vegetarian, it was not without regret. He liked meat. He was leaving behind foods that had sustained and comforted him. He didn't know if he would ever enjoy them again.

One such food was hamburger. Hamburger is special. There's no artifice to hamburger. It's plain, honest food. It says, simply, "Protein." Less simply but more precisely, "Protein, fiber, water, and fat, sandwiched between simple carbohydrates." That's it, no secrets, no surprises. And there's nothing between you and your meal—you pick it up and eat it with your hands.

Now, there were and there are vegetarian burgers. They, or the ones he tried, didn't satisfy the way a hot, savory, crunchy, meaty, juicy, hamburger satisfies. The challenge was to devise a satisfying vegetarian burger.

I accepted the challenge instantly. But it turned out to be incredibly difficult. It took me years to get anything remotely good.[1]

Let's start with some principles:

- A veggie burger should be simply and familiarly flavored.
  My quarrel with commercial products is that they pile on flavors—onions, garlic, tamari, olives, cumin, etc. Hamburgers aren't exotic. A vegetarian burger should be comfortable

---

[1] I was doing other things too.

the way a hamburger is comfortable. It should remind you of home.

And a burger should be bland. Hamburgers are like a painter's blank canvas. The consumer should flavor it, with tomato, mustard, pickles, etc.

- A veggie burger should have roughly the same "mouth feel" as a real hamburger.

  It shouldn't be starchy. Hamburgers aren't starchy, they're meaty. A good vegetarian burger can't be meaty, of course, but it should definitely be meatish.

- A veggie burger should be fairly easy to put together.

  It takes seconds to form a beef burger and slap it on the grill. One recipe for vegetarian burgers takes an hour and a half to prepare. If they're that difficult to prepare, you aren't going to make them often.

- Finally, a veggie burger should have enough structural integrity to sit on a barbecue grill three minutes without slipping through the grate.

At the start I had what I thought was a brilliant idea for the key ingredient. My idea was to use Grape-Nuts, the Post breakfast cereal. When wet, Grape-Nuts has some of the feel of hamburger. When dry, Grape-Nuts might pass for the crunchy outside of a charred burger. Would this work? How would it work?

I got something good right away. That fired me up. Then I had failure after failure! Only the memory of that early success kept me from giving up.

It took about five years. But in the end, a recipe emerged.

Butter Central to the problem and central to the solution is fat. There has to be fat. Butter is the obvious choice. An analysis of hamburger suggests the proper amount (about a tablespoon per burger). Some of this will run out, but some must be retained. How is that accomplished?

Mushrooms Raw mushrooms absorb fat. Cooked, they hang on to most of it. I grind the mushrooms. But something has to hold the mushrooms and the Grape-Nuts together. How is that accomplished?

Cheese Mozzarella becomes stringy when it melts. That's the glue. Warm it up, mix it in, and the burger will stay together.

Shamburgers

(For four burgers)

    4 Tb butter
    1 cup (4 oz) shredded mozzarella
    2 portobello mushrooms, food-processed into
        little bits
    1 cup Grape-Nuts
    1 clean, topless, empty 6-oz tuna can

You should have all the ingredients measured and ready to go. Follow the next instructions in their exact order:

1. Melt the butter. Stir and cook until the moisture has bubbled off and the solids have browned a little bit.
2. Add the mushrooms, mix quickly, and then *immediately*
3. Add the cheese, stirring until it melts, then *right away*
4. Turn off the heat, add the Grape-Nuts, and mix well.

Now use the tuna can to form the patties. Put a fourth of the mixture in the can, press it down evenly, invert the can and slap it down on a flat surface. Repeat with the rest of the mix. Let the patties cool before cooking.

Here's a comparison between shamburger and hamburger (uncooked) (data from USDA National Nutrient Database):

	1/4 lb 85% lean ground beef	one shamburger
water (gm)	74.26	49.5
calories	243	253
protein	21.1	7.74
fat	16.95	14.41
saturated fat	6.63	8.85
carbohydrates	0	26.11
fiber	0	3.15

It's missing protein, though. And it's a little starchy.

But it's *good*. It works. You can grill it over charcoal. It won't fall into the fire. It cooks quickly. The cheese gives it nice browning.

Smother it with caramelized onions. Salt it. Pepper it. Top it with fresh tomato and pickle. Drape it with ketchup.

The mix is amenable to other uses of ground meat. You can make meatballs and sausages. Just be careful about when in the process you add the extra ingredients and watch how you cook the final product.

A MATHEMATICAL CHALLENGE

A friend brought me a problem. He had been in bed all day with the flu. He spent some of that time asleep and some of that time staring at a wall covered with a dull sort of wallpaper.

He imagined a line starting in one corner, moving diagonally, and bouncing off the sides.

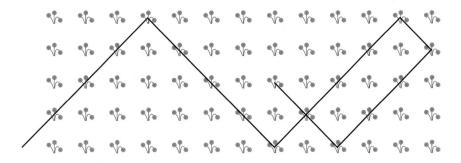

He followed this line until it hit a corner.

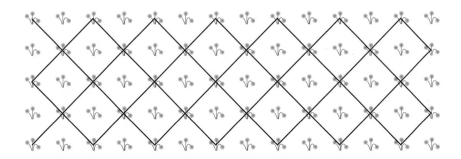

He thought that if the wall were smaller, the path would be different.

He checked this out by chopping off a column,

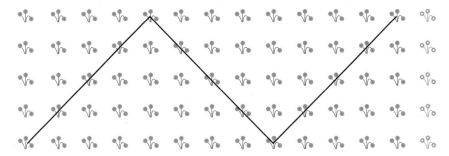

and by chopping off a column and a row.

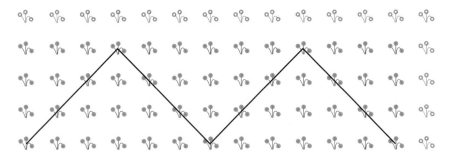

There was clearly a mystery here and he thought I might be interested.

I got very interested and we spent a few days puzzling it out. The problem is all about rectangles. You bounce around inside a rectangle. Do you bounce forever? If you don't, which corner do you hit?

We drew a lot of rectangles.

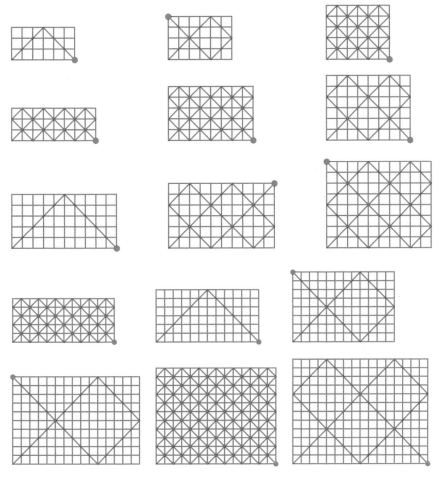

This has to make you curious!

We did figure it out, but it took us days. For a start, you don't bounce forever; you do have to end up at a corner. That's because you always go through cross points.

You can only go through a cross point twice

or an edge point once

so eventually you have to stop, and that has to be at a corner. And you can't end up at the starting corner because it's impossible to turn around.

Now to see *where* you end up, suppose we put dots on half the cross points, as on a checkerboard.

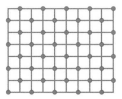

You can see that the path must always pass through dotted points.

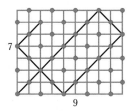

You can also see that if both side lengths are odd numbers, as in the case above, then you have to end up in the opposite corner—that's the only corner with a dot on it!

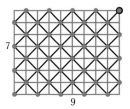

And if one side is even and the other is odd,

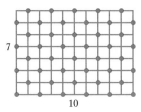

you have to end up on the even side. Again, that's the only corner with a dot.

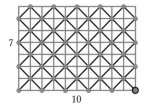

That leaves only the question of what happens when both sides are even, because in that case, all corners have dots.

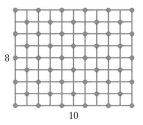

In that case, we just divide the side lengths by 2. The pattern with an 8 × 10 rectangle is exactly the same as the pattern with a 4 × 5 rectangle.

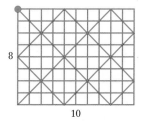

Or maybe it's clearer if I draw it this way:

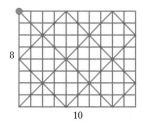
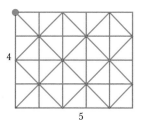

So that's the answer. The path always ends on the side that is "more even." You keep dividing by 2 until one side length is odd. Here's an example. Say the rectangle is 44 × 96.

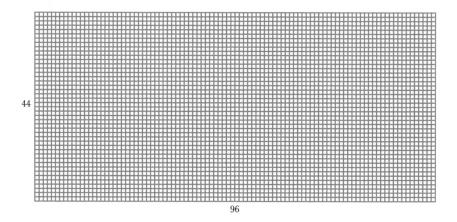

Both sides are even, so divide by 2.

$$22 \times 48.$$

They're still even, so divide by 2 again.

$$11 \times 24.$$

Aha. The 96 was more even than the 44 (we divided by 2 twice and it's still even). So the bouncing line will end up on the 96 side.

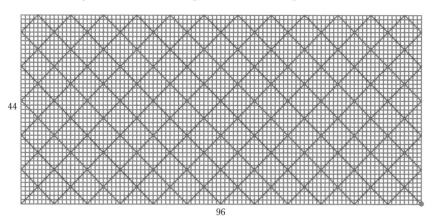

And if the two sides are equally even, you wind up at the diagonally opposite corner.

There's more to this story. I had a dream. It was a mathematical dream—I don't have many mathematical dreams, but this was one. I was sitting at a corner of a vast, but unknown rectangular polygon.

To explore it, I sent out a diagonal to see where it would go.

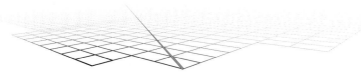

Then someone standing next to me said, "The shape of the polygon determines the pattern of diagonals. There's this amazing algebra . . ."

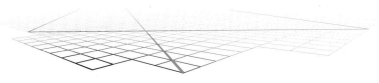

Then I woke up!

The idea stuck with me. Connected to this geometric . . . thing, there is this algebraic . . . thing and it explains . . . something.

Over the years since, I occasionally drew polygons and diagonals. I was looking for . . . well I didn't know what I was looking for. This was the only (mathematical) dream I had; I didn't have any other dreams to work on!

Then one day I had an idea. I'll explain what it was in a later chapter.

7

GLUTTONY

CURIOSITY ISN'T the only motivating force in mathematics and gastronomy—there's gluttony.

I'm a glutton. I eat enormous quantities, or I used to. I'm not big, though. I burn calories fidgeting.

My gluttony extends to mathematics. I want things and then I want more.

MORE SUDOKU

I don't really understand it myself, but the "Clueless Sudoku" project (in chapter 3) won't quit. Whenever I sit down with a piece of paper, I start drawing squares and dividing them into regions. In the dentist's office. At concerts, lectures. Watching the *Daily Show*. The people around me have been pretty tolerant.

My original group of collaborators found 7 × 7, 8 × 8, and 9 × 9 puzzles but I wasn't satisfied. The puzzles are rare and difficult to find. And when found, they're difficult to solve (the big ones) and not especially enjoyable.

Some new students[1] and I loosened the rules a bit. We allowed regions to connect by just a corner. You may recall that there are no clueless 4 × 4 puzzles with five regions. Now, allowing corner connections, we can find such puzzles.

---

[1] Sonia Brown, Bayla Weick, and Christine Niccoli.

There's also a puzzle with four regions.

There's even a (not very interesting) 2 × 2 puzzle.

We loosened things further by allowing blank squares, as in cross-word puzzles. Look at how much fun this can be. Here's a 4 × 4.

The numbers in a full 4 × 4 square add to 40, but in this square the sum will be less because there are two blanks. How much less? It's hard to say!

There are three regions, each with the same sum, so the total sum has to be a multiple of 3. It can't be 39 because the blanks cover up numbers adding to more than 1.

Can the total be 36? It can't. The most that the top left region can be is 11 (two 4's and one 3). So the total can't be higher than

$$3 \times 11 = 33.$$

So the missing numbers (where the blanks are) should add to 7, that is, they should be 3 and 4. That gives us

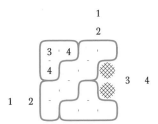

Now look at the middle region. It will have a 1 and a 2 on the top line and a 1 and a 2 in the left column. That adds up to 6, so in the middle there have to be numbers that add up to 5. There are two ways to do this.

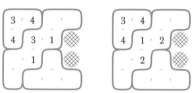

If we fill in the rest of the numbers where they have to go, we see that one of these works and the other doesn't.

I've got more.

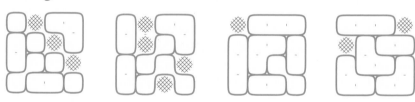

Now let's change things again. How about puzzles where the sums in the regions can be different, but they all have to be prime numbers?

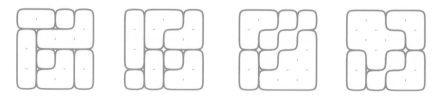

We're not the only ones inventing sudoku variants and other puzzles. Mathematician Laura Taalman and computer scientist Philip Riley have volumes of fascinating puzzles.[2]

---

[2] See Philip Riley and Laura Taalman, *Naked Sudoku* (Sterling Publishing, 2009).

APPLE PIE

The story of why I started cooking is not inspiring. My motives weren't pure. Indeed, they involved several important sins.

I really am a glutton. I love to eat. As a child, I ate well; my mother was a wonderful cook. But I always wanted more than I got, especially dessert. And of all desserts, it was apple pie I craved most. Not diner pies, not restaurant pies, and not bakery pies, but real, homemade apple pies.

When I was six, I had my first homemade apple pie. It was at my grandmother's house. I don't remember how it tasted, but I can still recall the gleam in my mother's eye when she explained the secret of the pie. "I watched her make it. Before she put on the top crust, she dotted the whole thing with big pats of butter!"

Several times as I was growing up, my mother made apple pie. Each one was a gem. But they were too few—only three or four before I went off to college. They were amazing pies. The apples were tart and sweet. Fresh fall apples, so flavorful no cinnamon was needed. The crust was golden, light and crisp, dry when it first hit the tongue, then dissolving into butter.

I grew up. I got married. I started a family. All the while, I longed for that pie. Eventually I set out to make one.

Success came pretty quickly, and it's not hard to see why. The fact is, despite apple pie's storied place in American culture, most apple pies sold in this country are abysmal. A pie of fresh, tart apples and a crust homemade with butter or lard, no matter how badly it's made, is guaranteed to surpass a commercial product.

That means that even if you've never made a pie before, you can't go seriously wrong. The chief difficulty is the crust, but I've developed a reliable method. Except for this method, the recipe below is standard.

Apple Pie

For the filling:

    5 cooking apples (yielding about 5 cups of pieces)
    1/4 to 1/3 cup sugar
    2 Tb butter
    1/2 to 1 tsp cinnamon
    lemon juice, if necessary
    1 tsp flour, maybe

For the crust:

    2 cups flour
    1 tsp salt
    2/3 cup lard or unsalted butter (1 1/3 sticks)
    water

The crust is crucial. I'll discuss its preparation last. Assume for now that you've rolled out the bottom crust and placed it in the pie pan.

Core, peel, and slice the apples. Place them in the crust. Sprinkle with sugar and cinnamon. Dot with butter. Roll out the top crust and place it on top. Seal the edge however you like. In about six places, jab a knife into the crust and twist to leave a hole for steam to escape. Sprinkle the crust with the teaspoon of sugar.

Bake in a preheated oven for 15 minutes at 450° and then another 35 minutes at 350°. Allow to cool. Serve, if you like, with vanilla ice cream or a good aged cheddar.

Now, the crust:

Mix the flour and salt in a large bowl. Place the lard or butter or lard/butter in the bowl. Cut it in with a pastry cutter.

Next, the water. Turn the cold water on in the kitchen sink so that it dribbles out in a tiny trickle. Hold the bowl with the flour mixture in one hand and a knife in the other. Let the water dribble into the bowl while you stir with the knife. The object is to add just enough water so that the dough is transformed into small dusty lumps. Don't be vigorous with the knife, but don't allow the water to pool. If the water is dribbling too fast, take the bowl away from the faucet from time to time. When you're done, the dough will still look pretty dry.

Recipes usually call for about 5 tablespoons of water. This method probably uses about that much.

Actually, the dough will look so dry that you'll think it won't stick together
when it's rolled out. In fact, it probably *won't* stick together, but trust me. This
is going to work.

Tear off a sheet of plastic wrap and lay it on the counter. Place a bit more
than half the dough on the sheet and cover it with a second sheet of plastic.
With a rolling pin, roll the dough out between the two sheets. Roll it roughly in
the shape of a rectangle.[3]

It won't look great and it probably would fall apart if you picked it up.

Don't pick it up. Remove the top sheet of plastic wrap and fold the bottom
third up, and fold the top third down,

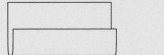   

then do the same horizontally, right and left.

Now replace the top sheet of plastic wrap and roll the dough out gently
into a disk.

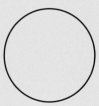

continued

[3] At this stage, my mother's crust usually looked, in her words, "like the British Isles."
You can do better, but this method is designed to deal with irregular results.

continued

This time it should look pretty decent. This time the dough will stick together. You should be able to remove the top sheet of plastic and, using the bottom sheet, turn it over into the pie pan. The crust should settle in nicely without breaking.

Form the top crust the same way.

This method rolls each crust twice—usually not a good idea because working the dough makes it tough. But remarkably, crusts produced this way are tender and light. I'm not sure why but I suspect it's because the dough is fairly dry.

Notes:

- Cooking apples are tart apples. The best I know is the Rhode Island Greening, but they're hard to find. Baldwins and Jonathans are decent, but they're hard to find too. The British Bramleys are terrific. I've made good pies from the French Calville Blanc d'Hiver. But we're not living in good apple times. Most stores don't sell apples for cooking. When in doubt, use a mixture.
- The lemon juice and the larger quantity of cinnamon are for when you have tired apples with no oomph. The cheese also serves this purpose. It should be a respectable old cheddar and it should be at room temperature.
- Consumption of too many commercial pies makes me loath to add flour or cornstarch to pie filling. The flour is here in case you fear your apples will be too juicy. I don't mind juice in a pie, in moderation. If adding flour, mix the apples, sugar, cinnamon, and flour in a bowl before pouring into the crust.
- Lard is best. Its melting point is higher than butter's. It successfully separates the flour into layers for a light, crispy crust. Butter is more likely to saturate the flour and produce a heavy crust. Some like half butter/half lard, preferring butter for its flavor. But the flavor of lard is nice too, and its porkiness is wonderful with apple.[4]

---

[4] What about using bacon fat? That's pretty porky. But don't use bacon fat. I've tried bacon fat. Its melting point is even lower than the melting point of butter (I think). The crust it makes isn't good. It is, though, unforgettable.

You haven't said anything, but I know what you're thinking. There's supposed to be mathematics in this book and practically everything you've seen is either a doodle, a puzzle, or a game. You want to know when I'm going to get serious.

I have two answers to that.

The first answer is that I'm *not* going to get serious. I'm just not a serious guy. If you want a serious book, there are many to choose from. I have a list somewhere.

The second answer is that I *am* serious. Doodles, puzzles, and games are genuine mathematics. Despite their outward frivolity, they contain ideas that are subtle and deep. Indeed, some philosophers have argued that *all* mathematics—when context is cleared away—consists of doodles, puzzles, and games.

And I want to add that what I'm saying about the parallels between mathematics and cooking is equally serious. The analogies I draw reveal aspects of both that are significant and have important consequences.

Of course I don't claim that everything in this book is profound. Every now and then something silly seems to slip in.

8

VANITY, SLOTH, PARSIMONY, AND LUST

IT'S A slippery slope. Having started on gluttony, we descend quickly into other sins.

VANITY

If you're good, you want to show off. If you like showing off, you get good so you can show off. Vanity is inextricably linked to self-improvement.

I am not innocent of this vice. I once memorized the squares of all two-digit numbers. I did it so that if someone said, "What's the square root of 5,329?" I could instantly answer "It's 73!" and everyone would be amazed.[1]

I also like to give dinners and hear people praise my dishes. But I strive against vanity! I am assisted in this by . . .

SLOTH

Why work to improve myself if that's vanity? No! I'll rest. I'll read comics. It's much more relaxing to be virtuous.

This seems to be a serious argument against improving oneself:

> Self-improvement is vanity.
> Vanity is a sin.
> ∴ Self-improvement is sinful.

---

[1] It didn't last. I forgot most of them in a few days.

I think there's a flaw here. But I just don't have the energy to look for it.

CULINARY SLOTH

The training of cooks covers many methods of avoiding work. Quick ways of breaking eggs, slicing vegetables, etc. Consider the beautiful method of making puff pastry.

To make puff pastry, you start with a single layer of butter surrounded by dough.

The combination is then rolled out

and folded in three,

creating three layers of butter and thus, four layers of dough.

The folding is done three more times. Each time the number of layers of butter is tripled: 9 layers, 27 layers, 81 layers. This creates, successively, 10, 28, and 82 layers of dough.

Eighty-two layers of dough with just four operations.

Sloth in action.

MATHEMATICAL SLOTH

Mathematicians are masters of sloth. A single theorem can prove infinitely many facts. Here's an example. In chapter 6, we studied the path of a point bouncing in a rectangle.

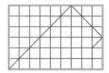

Let's find the length of the path, not just for one rectangle, but for all rectangles. The length of the path in the example above (a 6 × 9 rectangle) is 18 (as measured by the number of squares the path passes through).

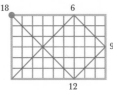

In this case, the length is the least common multiple of the side lengths, 6 and 9. But it turns out that the length is always the least common multiple of the side lengths, no matter what the (integer) side lengths are. If we had to show this for each rectangle one-by-one, we would have to draw infinitely many pictures. Instead we'll use a single argument. Sloth in action!

The key is to look at the bouncing point in two different ways. On the left we'll just bounce. On the right, we will pass through the side of the rectangle into its reflection.

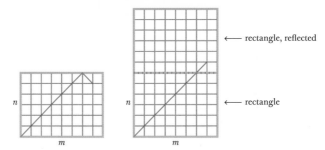

And every time we reach a side, we bounce in the left-hand picture and we pass into a reflection in the right-hand picture.

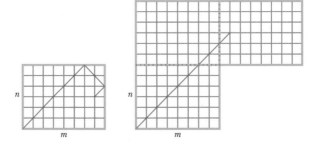

As we do this, the path on the right is a straight line.

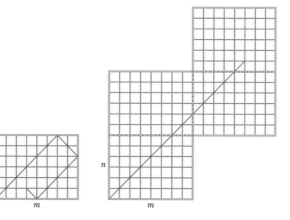

And when we're done,

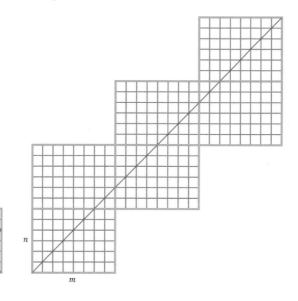

we see that the diagram at the right is contained in a square. The side of the square is simultaneously a multiple of the rectangle's width and a multiple of its length.

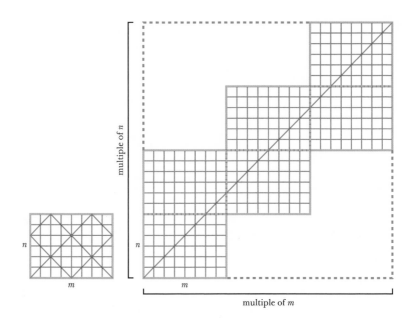

It's the least common multiple of the side lengths, the first time that the diagonal hits a corner.

JUST A LITTLE MORE SLOTH

There's another benefit to sloth.

I play squash. Some of the people I play with take lessons to improve their game. I don't. I don't work to improve my play. But since I play once or twice a week, I do improve—but very slowly. The result is that I'm peaking now, in my sixties. Had I really worked at my game, I probably would have peaked in my forties. I would be a sour old man now, raging at my decline and boring people with stories of my glorious past. Instead I annoy opponents by playing slightly better than they expect.

There's a lesson here for both mathematics and gastronomy. If you spend time on something, you will improve. Experiment in the kitchen, play with ingredients. You'll learn to cook well. Play with puzzles and games. Your ability to solve math problems of any sort will grow.

There's an old joke about mathematicians that gets to the essence of mathematical sloth.

Consider the following problem: There are two buckets, one white and one red. The white bucket is filled with water. The red bucket is empty. There is a small fire.

We ask the mathematician, "How can we put out the fire?"

The mathematician replies, "Pour the water in the white bucket over the fire."

Now suppose the situation is slightly different. Suppose that the white bucket is empty and the red bucket is filled with water.

Again we ask the mathematician, "How can we put out the fire?"

The mathematician replies, "Pour the water in the red bucket into the white bucket, thus reducing the problem to the earlier problem which we've already solved."

Most mathematicians, but not all, get this joke.

PARSIMONY

Parsimony may be a sin, but it's a motivating factor in both cooking and mathematics.

In mathematics, it's a virtue to do more with less. More than two thousand years ago, Euclid reduced all of mathematics to five postulates. From five assumptions, Euclid derived all the mathematical results known at that time.

Marvelous as that achievement was, some wondered if even more stinginess could be achieved. Was it possible that not all five assumptions were needed? In particular, could the fifth postulate be derived from the first four?

This question motivated much work in mathematics. It wasn't until the nineteenth century that it was proved that the answer was no. The fifth postulate can't be proved from the other four.

GASTRONOMIC PARSIMONY

Let's return to my youth. I was in my twenties. I was a graduate student of limited means. I loved pizza but pizza was expensive in those days. Pizza had a different status then. There were no nation-wide chains. You got pizza at funky, exotic joints. I grew up looking at pizza as older-brother-food, something the big kids got.

I felt that way well into my thirties.

To have pizza without paying for it—that was my goal. It sustained many years of pizza attempts. Even then (sloth) I didn't read up on pizza. I just tried this and that, expecting lightning to strike. It probably took me twenty years to come up with a decent product.

> ### Pizza, basic, 16 inches
> I could give you here the recipe I developed, but I won't. There's a better one in chapter 12.
> Someday, you'll thank me for this.

LUST

I won't say much about lust. I'll simply give you two lines from George Bernard Shaw, one of the great intellects of the twentieth century.

There is no sincerer love than the love of food.

and

Sex is infinitely less interesting than mathematics.

Well okay, I'll say a little more. I'll tell you a joke:[2]

---

[2] Compliments to Arthur Apter, who told me this joke and couched it carefully in gender-neutral terms.

A mathematician was trying to decide: Should I get married? Or should I take a lover? The mathematician consulted a lawyer.

"By all means take a lover. The legal complications of marriage are immense. You're much better off with a simple affair."

The mathematician then consulted a doctor.

"By all means get married. Marriage is much healthier. Married people live longer. Don't distress yourself with the uncertainties of affairs."

Finally, the mathematician consulted another mathematician.

"Do both. Your spouse will think you're with your lover, your lover will think you're with your spouse, and you can do mathematics."

I have more sins, but confessions (except one's own) make dull reading. Instead we move on to a concept common to both cooking and mathematics, the concept of "recipe."

9

ON THE EDGE, AND OVER

MANY COOKS, probably most cooks, follow recipes. So do mathematicians. We call them algorithms, rules, procedures, or techniques. Sometimes we even call them recipes.[1]

If we think of a mathematical theorem as being like a dish, then a proof is also a recipe. The proof explains, in a series of careful steps, how you can reach the conclusion of the theorem from the premises.

But the excitement of mathematics and the thrills in the kitchen lie in creating something new. This is what high-end practitioners do. But anyone can do it. You just find the edge and jump off.

SOUNDS LIKE A RECIPE FOR DISASTER!

It is a recipe for disaster, and I strongly recommend you try it.

What stops many from going beyond recipes is that they don't know what will happen. "The recipe calls for flour. I really don't know what will happen if I use canned artichoke hearts instead!"

Indeed, and I don't either!

But you and I will never find out unless we try. Good cooks often don't know what will happen when they try something peculiar. They try it anyway because they need to find out. To make progress, you have to make experiments. *And you have to be willing to fail.*

COOKING OVER THE EDGE

We'll start with cooking because most home cooks have wandered away from a recipe. Maybe you were missing an ingredient. Maybe

[1] See, for example, William H. Press, Saul A. Teukolsky, William T. Vetterling and Brian P. Flannery, *Numerical Recipes: The Art of Computer Programming*, 3rd ed. (Cambridge University Press, 2007).

you couldn't find the recipe. Maybe you had something you wanted to use up. Experiments are fun! Failure happens. You get used to it.

There is a lovely recipe in an old *Joy of Cooking* for cornmeal pancakes. Cornmeal makes cornmeal mush, a hot cereal. The pancake recipe uses that mush. Well, farina is also a hot cereal. I like farina. One day I thought I would try using farina instead of cornmeal. It was an experiment that worked.

Cream of Pancakes

1/2 cup farina (also marketed
   as Cream of Wheat)
1/4 to 1/2 tsp salt
1 cup boiling water
1 egg
3/4 cup buttermilk
1/2 cup all-purpose flour
1 to 2 tsp baking powder
1/2 tsp baking soda
butter

Mix the salt and farina, pour in the boiling water, and stir vigorously with a whip to break up any lumps. Cover and let sit while you complete the preparations. It should sit at least 10 minutes.

Sift together the flour, baking powder, and baking soda.

Add the buttermilk to the farina a tablespoon at a time at first, mixing well (this is to make sure there are not farina lumps in the final product). Beat in the egg.

Dump in the flour mixture and stir together roughly (lumps are allowed now).

Heat a griddle. When hot, grease the pan liberally with butter and spoon out batter for pancakes about 4 inches in diameter. When the edges start to dry out, turn the pancakes and place a bit of butter on each. They'll be done less than a minute after turning.

continued

> *continued*
>
> *Notes:*
> - The batter should be runny. That makes the pancakes light.
> - I put no butter in the batter. My feeling is that butter is much nicer on the outside, where it hits the tongue.
> - The recipe for cornmeal pancakes uses milk and no baking soda. This has a better flavor and it removes the slightly chemical taste of baking powder.

Was that edgy? Maybe not. So here's another idea: blue cheese ice cream.

I don't remember how I first thought of it. The idea came. The act followed. My niece will never forget the day I made blue cheese ice cream. She won't let me forget, either.

I liked the flavor but it was sort of gritty.

Then recently, while thinking about baked Brie, the idea came back.

My thoughts went more or less like this:

Baked Brie? . . . That's really a dessert. . . . It's a sticky, gooey dessert. . . . And you can't taste the brie with that jam. . . . It's cheese abuse. . . . Who is running this country, anyway?

But a dessert with Brie might work. . . . What about Brie ice cream? . . . But freezing dulls flavors. . . . You need a stronger cheese. . . . Ah! Stinky cheese ice cream!

> ## Stinky Cheese Ice Cream
> 1/2 lb Cambozola cheese
> 1 cup milk
> 1/3 cup sugar
> 1 cup cream
>
> Cambozola is perfect. It's only slightly more pungent than Camembert. "Stinky" is great for shock value, but the ice cream really isn't stinky; funky, maybe, but not stinky.[2]

[2] Make sure the Cambozola is fresh. I did this once with an elderly Cambozola. It was several shades too funky.

Remove and discard the rind of the cheese. Put the cheese in a saucepan with the milk and sugar. Heat the mixture, stirring, until smooth. Turn off the heat. You will see, at this point, small strands of blue mold. That's good for shocking diners, but if you don't like it, run the mix through a blender.

Now add the cream, chill, and then freeze using your favorite ice cream device. You can actually freeze it without a device. I once poured the mixture into a plastic tub and put it in the freezer. I took it out every 10 or 15 minutes and stirred it with a whisk. It came out pretty creamy, especially when eaten right away.

That's edgy, but there's more. It seemed to me that this was the perfect vehicle for a caramel sauce. Caramel sauce has a problem—it's excessively sweet. But what if it were poured over a funky, salty ice cream?

Stinky Cheese Ice Cream with Caramel Sauce

one recipe stinky cheese ice cream
1 cup sugar
2 to 3 Tb water
1/4 tsp lemon juice
2 pinches salt
3/4 cup heavy cream

Make the ice cream.

Place the sugar, water, and lemon juice in a saucepan. Bring it to a boil over medium heat. Cook it until the sugar caramelizes. It should be a nice, not-too-brown brown. Turn off the heat. Slowly add the cream, stirring all the time, and maybe wearing mitts because there will be spattering. Add the salt. Cook for a couple of minutes until the sauce is totally smooth.

Spoon the sauce on the ice cream and enjoy.

MATH OVER THE EDGE

Where is the edge? How do you get there? And how do you jump off?

There are edges everywhere and much closer than you can imagine. My friend Tom Wiener, who teaches sixth grade, used to introduce his class to mathematical research each year. He spun them a yarn about a famous mathematician, Professor Etoile, who had devised a new way to write numbers. He didn't use the digits 0, 1, 2, 3, 4, 5, 6, 7, 8, 9. Instead he used the letters A, B, C, D, O. But Etoile, Tom told his class, fell off a boat and was eaten by an octopus before he could tell the world of his discovery. Tom challenged his class to devise a numeration system using the five letters A, B, C, D, O.

Each year, the class, working in separate groups, came up with systems. They created mathematics. Most of the systems looked a lot like Roman numerals. A typical system might use A to represent 1, B to represent 10, C to represent 100, D to represent 1,000, and O to represent 10,000. Then they would write 357 as

CCCBBBBBAAAAAAA

Some of the groups came up with pretty sophisticated ideas. Occasionally, students dreamed up systems resembling base 5—more about that in a moment.

The edge (in this case) is numeration. Getting to the edge is just noticing that you can invent your own system for representing numbers. Jumping off is accomplished by fiddling around and experimenting with ideas. Nobody gets hurt when you jump. All you can lose is time.

What you invent may be useless. It may even be annoying. But it will be your own personal system for representing numbers. You'll certainly have learned something by trying. And your system may have elegance and charm.

I was playing with the binary numeration system a few years ago and I thought of a possible variation.

For those not familiar with binary, it's just like our ordinary base 10 system but stripped down. In our usual system we have a 1's column, a 10's column, a 100's column, and so on.

...	1,000,000	100,000	10,000	1,000	100	10	1

That is,

...	10^6	10^5	10^4	10^3	10^2	10	1

We use the digits 0, 1, 2, 3, 4, 5, 6, 7, 8, 9. When we write 537, for example,

...					5	3	7
	10^6	10^5	10^4	10^3	10^2	10	1

We just mean five 10^2's, three 10's, and seven 1's.
In binary, or base 2, we replace the 10's with 2's.

...	2^6	2^5	2^4	2^3	2^2	2	1

That is,

...	64	32	16	8	4	2	1

And we use only two digits, 0 and 1. If we write

$$1,101$$

in binary, it means

...			1	1	0	1
	64	32	16	8	4	2	1

in other words, the number 8 + 4 + 1 = 13.

We only use 0's and 1's so that every number can be written in only one way. If we allowed 2's, for example, then 2 could be written both as 2 and as 10.

Surprisingly, with just 0's and 1's every natural number can be represented. To represent 43, for example, you think:

Well, I don't need a 64. I can use a 32.

...		1					
	64	32	16	8	4	2	1

If I use 32, then I need exactly 43 − 32 = 11 more. For 11, I don't need a 16, but I can use an 8.

	1	0	1			
...						
64	32	16	8	4	2	1

Now I need 11 - 8 = 3 more. I can get that with a 2 and a 1.

	1	0	1	0	1	1
...						
64	32	16	8	4	2	1

and that's how we represent 43 in binary:

101011.

Now, what I wondered was: *Do you have to use 0 and 1? Could you use, say, 1 and 2?*

I tried it and it worked. Here, for example, is how you represent 8:

				1	1	2
...						
64	32	16	8	4	2	1

And here's how to represent 43:

		1	2	2	1	1
...						
64	32	16	8	4	2	1

And here's how to represent the first twelve numbers:

ordinary base 10	1	2	3	4	5	6	7	8	9	10	11	12	...
binary using 1 and 2	1	2	11	12	21	22	111	112	121	122	211	212	...

There is one number you can't represent, though. You can't represent zero.

That was fun. I tried other combinations, 1 and 3, 2 and 3, 0 and 2, etc., but they all had problems. With 1 and 3, you only can represent odd numbers. With 0 and 2, you only can represent even numbers. With 2 and 3, there are gaps. You can't represent 1. You can represent 2 and 3 but you miss 4 and 5. You can get 6, 7, 8, and 9, but then you miss 10, 11, 12 and 13.

I was about to give up when I thought of ternary (base 3). Ternary uses these columns:

$$\ldots \quad \overline{} \quad \overline{} \quad \overline{} \quad \overline{} \quad \overline{} \quad \overline{} \quad \overline{}$$
$$\quad\;\; 3^6 \qquad 3^5 \qquad 3^4 \qquad 3^3 \qquad 3^2 \qquad 3 \qquad 1$$

That is,

$$\ldots \quad \overline{} \quad \overline{} \quad \overline{} \quad \overline{} \quad \overline{} \quad \overline{} \quad \overline{}$$
$$\quad\;\; 729 \qquad 243 \qquad 81 \qquad 27 \qquad 9 \qquad 3 \qquad 1$$

Normally in base 3 you use 0, 1, and 2. But I found that 1, 2, and 3 also work—except that again you can't represent zero.

Something else that works is 0, 1, −1. This system has a name, "balanced ternary." Not only can you write 0, 1, 2, 3, 4, . . . , but you can write all negative numbers as well, which makes the system *very cool*.

Are there other combinations? I tried a few. Here's one that worked, although it didn't look promising at first: −1, 1, and 3. I call it "unbalanced ternary." For simplicity, I'll write **1** for 1 and $\bar{1}$ for −1. It's easy to see how you write 1:

1.

What about 2? It's like this:

1$\bar{\mathbf{1}}$,

plus a 3 and minus a 1 equals 2. Then 3 and 4 are easy:

3 and **31**,

but 5 had me stumped until I found

1$\bar{\mathbf{1}}\bar{\mathbf{1}}$.

Here's how you write the first twelve numbers:

ordinary base 10	1	2	3	4	5	6	7	8	9	10	11	12	...
unbalanced ternary	1	1$\bar{1}$	3	11	1$\bar{1}\bar{1}$	13	1$\bar{1}$1	3$\bar{1}$	1$\bar{1}$3	31	11$\bar{1}$	33	...

You can get 0.

1$\bar{\mathbf{3}}$.

And you can get negative numbers. I'll let you fill this in:[3]

ordinary base 10	−1	−2	−3	−4	−5	−6	−7	−8	−9	−10	−11	−12	...
unbalanced ternary	1	11	113										...

As I mentioned earlier, sometimes Tom's students created systems that looked like base 5. Base 5 has these columns:

$$\cdots \quad \overline{5^6} \quad \overline{5^5} \quad \overline{5^4} \quad \overline{5^3} \quad \overline{5^2} \quad \overline{5} \quad \overline{1}$$

and it uses the digits 0, 1, 2, 3, 4. Tom's students probably did something like use A to represent 1, B to represent 2, C to represent 3, D to represent 4, and O to represent 0. Then

BDC,

for example, would name $2 \cdot 5^2 + 4 \cdot 5 + 3 = 50 + 20 + 3 = 73$.

I wonder, what sets of five numbers could be used as digits with those columns? I'm pretty sure 1, 2, 3, 4, 5 works, but what else?

There's only one way to find out!

"Edge" has an intimidating sound (in mathematics, anyway). But you don't have to look far to find it. And it's safe. Sixth graders can play on it. Nobody gets hurt.

---

[3] On the website is a proof that every number can be represented. Oddly, every number can be represented infinitely many different ways. That's also on the website.

10

THINKING GLOBALLY

YOU CAN live close to the edge and not see it. But there are highways, in both mathematics and gastronomy, that lead directly to it. The simplest of these is generalization. Often we can advance by taking something successful and pushing it to see if it will go further. Sometimes it does.

THE ALL-PURPOSE DOUGH

When I finalized my bread recipe (chapter 2), I wanted to use it everywhere. I did successfully use it to make pizza. But then I wanted to see if it would work in any recipe where a yeast dough is needed—sticky buns, bagels, English muffins, etc. I found that it was serviceable in surprising ways. That meant that I could make dough for a couple of loaves, bake a loaf, and put the rest of the dough in the refrigerator where it was ready for any of several occasions.

Here are some of the most successful uses of the dough.

Sticky Buns

1/2 recipe bread dough (one loaf 's worth)

5 to 8 Tb unsalted butter

pecans

2/3 cup sugar

a little molasses

raisins (I like golden raisins)

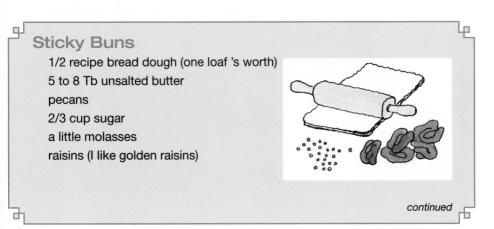

continued

continued

Grease a pan with butter. I prefer to use something at least 75 square inches in area with sides 2 inches high. I used to use a square 8 × 8 inch pan and the butter would always flow over the sides and burn at the bottom of the oven. It gave the buns a distinctive taste. After a few years of this I switched to a larger pan.

Dot the bottom of the pan with 2–3 Tb of the butter. Sprinkle pecans in the pan, as many or as few as you like. Sprinkle 1/3 cup of the sugar in the pan. Dribble a little molasses over the sugar (a teaspoon or two).

Roll out the dough in a large rectangle.

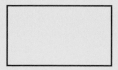

Dot with the remaining butter, sprinkle the remaining sugar, dribble some more molasses, and sprinkle on the raisins (as many as you like).

Roll up the dough starting with the long edge.

Slice the roll into 16 pieces and place them cut-side up (or down) in the pan.

Let the buns rise until they've puffed a bit, then bake at 450° for 25–30 minutes.[1] How long you bake depends a lot on the pan you use.

Notes:

• You should be able to avoid undercooking. Press in the middle. If it's spongy, it should cook longer.

[1] That's right, 450°.

- You may be able to avoid burning. If you're using a glass pan, look at the bottom.
- But my experience is that this stuff is wonderful no matter what. Sometimes I burn it. It's great! Sometimes I take it out too soon and it's a little goopy in the middle. It's great!
- These buns aren't as sweet as commercial buns. They aren't overly greasy either. They have a nice crunch; it takes a little muscle to eat them. This is good—after the effort of eating a few, you feel you've earned a few more.
- You can vary the nut and the fruit. Almonds (sliced) and dried cherries is a lovely combination. Before rolling, sprinkle on 1/4 teaspoon of almond extract.
- Another excellent combination is chopped hazelnuts and dried apricots.

Baguettes

1 recipe bread dough
butter
cornmeal

Just make long, thin loaves. You can bake these on heavy-duty cooking sheets. Grease the sheets with butter and sprinkle with cornmeal. I divide a loaf 's worth into three or four baguettes, stretching them out anywhere from a foot to 20 inches.

Bake at 450° until you think they're done (20–30 min.).

At dinner parties, I make a loaf for every guest.

Recently I bought long pans, each designed for a pair of baguettes. They had cross sections looking like this:

Then I bent them so now they look like this:

continued

continued

I find that I don't have to grease these pans. When the bread is done, the loaves slide out easily.

Pita Bread
(For about eight pitas)

1/2 recipe bread dough
1 gas range

Put a griddle or frying pan on fairly high heat. Divide the dough into eight lumps. On a floured board, roll each lump out to a thickness of about 1/8 inch. When the griddle is hot, reduce the heat to moderate and slap on a disk of dough. When you see bubbles begin to rise on the surface (after a minute or two) flip the pita.

When the pita has stiffened slightly (when it doesn't flop much when an edge is lifted), flip it over, remove the griddle, and place the pita directly on the burner over the flame. It will soon begin to puff up. I keep the flipper hovering over the pita so that it doesn't puff too quickly in one spot (causing a rupture). When puffed and slightly charred, turn it over to brown the other side. I rather like it burned a little on both sides (not everyone does).

Cool on a rack. The cooled pitas can be cut in half and each side pried open to produce a pocket. These are best eaten within a half-hour.

Notes:
- They don't always puff. Why not? Many reasons, probably. Keep trying. Pitas that don't puff can still be opened as pockets. It takes more work, but they're delicious.
- I've tried this on an electric range. It didn't work. But maybe somebody smarter than me can succeed.

Some uses were not great.[2] The bagels were okay. The calzones were good. The English muffins were good but unusual. The strudel did not work out. But the pretzels were great. And here's one more:

---

[2] The experiment grouting tiles was a total failure.

Chinese Meat Buns

1/2 recipe bread dough

tasty filling

butter

wax paper

Divide the dough into 8–10 pieces. Roll each piece out and place some tasty filling in the middle. Bring the dough up around the filling on all sides and seal. Traditionally this is done by twisting the dough at the top. Place each bun on a square of buttered wax paper.

Steam the buns in a steamer for about 20 minutes.

For the fillings, consult Chinese cookbooks; barbecued pork is traditional. Domestic pulled pork with barbecue sauce is excellent. Philippine *adobo* works well. In the Philippines the buns are called *siopao*. Choose a filling recipe that isn't too soupy.

BOXES

Having pushed the bread dough along, we're now going to shove the rectangle.

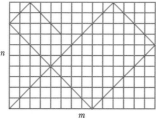

You may recall the dream at the end of chapter 6. It was of a mathematical theory connecting shapes (like the rectangle) and patterns of bouncing points. It was pretty vague.

My first attempt to bring the dream to reality was to look at ell-shaped regions.

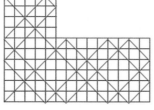

This quickly got very complicated, so I put it aside.

Moving in a different direction, I thought about going three-dimensional. I thought about a point bouncing around inside a box.

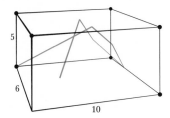

This turned out to be too easy. The answer was essentially the same as for the rectangle. In the rectangle, you end up along the side that is most even (has the most factors of 2). That's true here as well. In the box above, 6 and 10 are equally even and more even than 5, so you end up at the corner along both those sides.

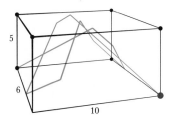

It's just a matter of the evenness of the dimensions.

Out of curiosity, I wondered what would happen if instead of bouncing around inside the box, you wound around the outside of the box like a ribbon. Let's say you start along the bottom.

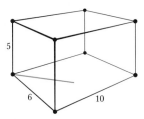

When you hit an edge, you just bend around up the side, keeping the same angle (45 degrees),

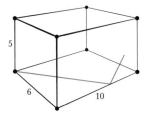

winding around at every edge, until something happens.

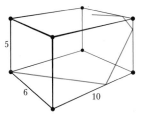

In this box, there's a surprise: you end up where you started!

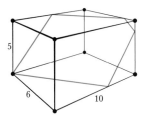

I found that boxes are at least as tricky as ells, but they fascinated me. As with rectangles, you can prove that all paths do end at corners, if the dimensions of the box are whole numbers. The reason is the same. A path always goes through cross points

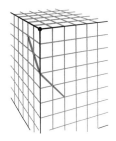

and it can only go through the same cross point twice.

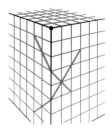

So a path can't go on forever. Eventually it ends up at a corner. But which corner?

That's a puzzle! It certainly depends on the side lengths of the box, but how? Evenness does matter, but it's not simple. Look at these, for example.

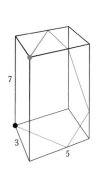 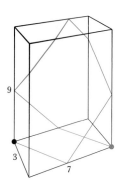 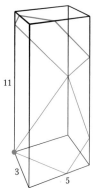

All the dimensions are odd, but there's a different ending point for each box!

I've been working on this with my son, Fred. We've made considerable progress. Some of our work is on the website. More is coming in a later chapter.

11

EATING LOCALLY

WHAT, IN mathematics, constitutes "local ingredients"? Nothing that I can think of. The elements of mathematics have no location on Earth; they exist in our minds. This may be a feature where gastronomy and mathematics are basically different.

There are two reasons for using local ingredients. One is that dishes are vastly better when made with fresh components. The other is that the preparation of meals has a much lower impact on the environment if food is not shipped thousands of miles. But of course, mathematics suffers from no problems of this sort.

Indeed, mathematics is totally green. We never throw a number away. We reuse numbers constantly. Mathematics, I'm happy to say, is completely sustainable.

INGREDIENTS WITH INGREDIENTS

If you don't use local ingredients, then your dish is one more step removed from its origins. The same is true if you use ingredients that have ingredients. A recipe might call for a tablespoon of ketchup, for example. But what is "ketchup"? Depending on the manufacturer, it might contain tomato concentrate from red ripe tomatoes, distilled vinegar, high fructose corn syrup, corn syrup, salt, spice, onion powder, and natural flavoring.[1] Are you cooking this dish or outsourcing it?

Most cooks don't mind using ingredients with ingredients. They may prefer one brand of ketchup to another, but they'll use ketchup. At the high end, though, a chef may want to break this down, tease

---

[1] Heinz Tomato Ketchup, for example. I spend a lot of time in the grocery store reading labels.

out what she or he really wants from ketchup, and then supply it from basic ingredients.

I'm not a high-end chef. But I am bothered by ingredients with ingredients. I sometimes make things so that I don't have to use commercial products. I once made ketchup (there's a recipe in *Joy of Cooking*).[2]

I routinely make my own brown sugar (just mix white sugar and molasses). Troubled, at one point, by my reliance on Grape-Nuts in the recipe for shamburgers (chapter 6), I devised a recipe for Grape-Nuts.

But then I wondered: *Why am I bothered by ingredients with ingredients?*

PROOFS WITH PROOFS

And then I had the answer. Suppose that I'm reading a math paper and in that paper I read the proof of a theorem. Then I'm uneasy if the proof uses another result, a proposition from a different paper. I feel that I don't really understand the proof of the theorem until I've read and understood the proof of the earlier proposition. I feel this especially keenly as a mathematical logician.

Mathematical logic is the first area I worked in seriously. Logic is concerned with the foundations of mathematics. We start with a few axioms and from these build the great edifice of mathematics. That's what Euclid did, twenty-four hundred years ago. That's what logic did again in the twentieth century.

I was brought up to look for first principles, to ground my work on mathematical bedrock. Not all mathematicians feel this way; in fact most don't. We wouldn't make much progress if everything had to be done from first principles. Perhaps this is another aesthetic— both of mathematics and of gastronomy. It's not surprising, then, that I embrace that aesthetic in both disciplines.

---

[2] Irma S. Rombauer and Marion Rombauer Becker (Bobbs-Merrill, 1971), and a more recent one by Melissa Clark, *The New York Times Magazine*, June 29, 2012.

A RECIPE WITH HIDDEN INGREDIENTS

Most cooks are comfortable with composed ingredients. That's a style too. In fact, cooking with commercial products was big when I was a kid: cookies made with corn flakes, casseroles made with canned mushroom soup, meatloaf made with bottled spaghetti sauce, etc.

My favorite convenience dish was the clam dip that was probably an invention of Kraft Foods. My mother made it for parties. On the morning after one of those parties, my brothers and I would scoop up the remains with leftover Fritos. I fell in love.

I went on the web recently looking for a recipe. I found many, but none that matched my memory—a can of minced clams, a couple of packages of cream cheese with chives, and some Worcestershire sauce. I can no longer find the 3-ounce cream cheese with chives. The cream cheese with chives and onion I've found is gooey; perhaps it has less cheese and more water. Given all that, the recipe below is close to what I remember.

Clam Dip

6.5-oz can of minced clams
6 oz cream cheese
1 1/2 tsp Worcestershire
     sauce
1 1/2 tsp chopped chives

Reserve the liquid in the can. Mix the rest of the ingredients together. Add enough of the reserved liquid to make a good dipping consistency. Chill. Serve with Original Fritos, the real thing.

A PROOF WITH A HIDDEN PROOF

It's not just logicians who are bothered by proofs with proofs. There comes a time in most students' lives when their teacher tells them (with a mixture of pride and dread) that the infinite decimal

$$.999999\ldots$$

equals 1. Most students don't buy it. There's a fight. The teacher is the winner, but victory is hollow, because in their hearts the students are not convinced. Here's how the proof usually goes:

Alright, we have this decimal. What is it? Let's say we don't know what it is so we call it x.

$$x = .999999\ldots$$

Now we multiply both sides of the equation by 10. On the right, that means just moving the decimal point.

$$10x = 9.999999\ldots$$

Now we have two equations so we can subtract:

$$
\begin{aligned}
10x &= 9.999999\ldots \\
- x &= .999999\ldots \\
\hline
9x &= 9
\end{aligned}
$$

Well! If $9x$ equals 9, then there's only one possible value for x:

$$x = 1 \ !$$

Victory is hollow because this marvelous proof convinces no one, perhaps not even the teacher.

The students are thinking: "No matter how far out you go, .999999 . . . is less than 1. There's always a bit left over!"

That's a proof too, a proof that .999999 . . . doesn't equal 1. The problem is that students don't see .999999 . . . as a completed number. They see it as a moving object. It's not an infinite decimal, it's a finite decimal getting longer and longer. The finite decimals are never 1. So .999999 . . . doesn't equal 1.

What's missing is the entire theory of infinite sums, because .999999 . . . is the infinite sum

$$\frac{9}{10} + \frac{9}{100} + \frac{9}{1000} + \frac{9}{10000} + \ldots$$

Without the theory, the first proof is meaningless. To say "x = .999999 . . ." is to presume that .999999 . . . represents a single, unchanging quantity. Why should it represent a quantity?

The hidden theory is not simple. It's fair to say that it took the human race more than two thousand years to work out the details!

In the end, it all depends on what kind of mathematician/cook you are. I'm both, I suppose. Sometimes I long for first principles. I'll start with the empty set and create an artisanal set of numbers. I'll also make my own sour cream, wine vinegar, brown sugar, ketchup, mustard, and mayonnaise.

And at other times, I'll kick back with clam dip, take an infinite decimal off the shelf, and chill.

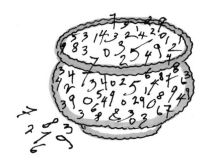

12

THE HUMBLE COOK

IN THE Arrogant Chef (chapter 2) I introduced my theory of problem-solving. I explained there that it takes two personalities to solve a problem, someone who is full of self-confidence and someone who is awash in self-doubt. We saw the first personality in that chapter. It's time to talk about the second.

But first let me tell you a true story of confidence and humility in the classroom. I heard this from a colleague some years ago. He had given a test. The class had not done very well but he was impressed with the paper one woman had turned in, the best in the class. He singled her out for public praise. Immediately, two men stood up to complain. There was something wrong with the test, they said. The woman, while smart, didn't understand the material as well as they did. In fact, they said, she had come to them for help the day before the test. She'd had difficulties and they'd helped her. Why was it that she scored higher on the test than they did?

The woman was embarrassed. The teacher was nonplussed and had trouble responding. But seen from the point of view of my theory of problem-solving, the explanation is clear. The woman had the humility to ask for help. The help she received gave her the confidence to go with the humility. The men had plenty of confidence, but in their exchange with the woman, they didn't gain any humility. Possibly, they even lost some!

It's not easy making the switch from confidence to diffidence. If you're sure you're right—why doubt yourself? Of course you *might* have made a careless mistake. The guys in the story must have done a few of those.

But let's suppose you're right, *actually right*. You *still* need to doubt!

Does that sound strange? I'll tell you two stories.

DOUBTING EULER

This story is told in rich detail by the philosopher Imre Lakatos (1922–1974) in *Proofs and Refutations*.[1] I'll need just a fragment of his discussion to make my point.

The story is about polyhedra.

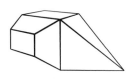

Think of a polyhedron as a solid figure with polygonal sides. The surface of the polyhedron consists of polygons (called "faces"), edges (the line segments between faces), and vertices.

The story starts with the observation by Leonhard Euler (1707–1783), one of history's greatest mathematicians, that in any polyhedron, the number of faces plus the number of vertices is two more than the number of edges.

$$F + V = E + 2,$$

where F is the number of faces, V is the number of vertices, and E is the number of edges. The cube, for example, has 6 faces, 8 vertices, and 12 edges.

$$6 + 8 = 12 + 2.$$

Proofs of Euler's formula were suggested, but in a few years mathematicians began to doubt it. They found examples where the formula didn't hold.

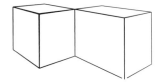

In this example, technically a single polyhedron, there are 12 faces, 14 vertices, and 23 edges.

$$12 + 14 \neq 23 + 2.$$

[1] Cambridge University Press, 1976.

This was addressed by restricting the formula to figures where each edge is attached to exactly two faces.

But here's another example.

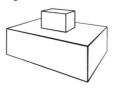

Now there are 11 faces, 16 vertices, and 24 edges.

$$11 + 16 \neq 24 + 2.$$

The problem is that one of the faces has a "hole."

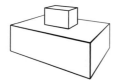 

The problem was fixed by requiring that the faces not have holes. If the face with a hole in the picture above is broken up,

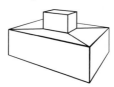

then the formula works.

$$14 + 16 = 28 + 2.$$

Good! But then there's this example.

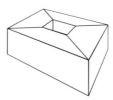

The faces don't have holes but the polyhedron has a hole so we have 16 faces, 16 vertices and 32 edges!

$$16 + 16 \neq 32 + 2.$$

Over and over the examples came. Each time, the statement of the theorem was refined. Always there was a true theorem at the center.

The period from Euler's first statement to the final, rigorous version with proof was almost fifty years. A more complete discussion is beyond the scope of this book.[2]

DOUBTING PIZZA

Over the years, the task that has humbled me the most is pizza. I started at the bottom. My first pizza crust was made with a baking powder dough (no yeast). It was so bad that humility trumped confidence. It was two years before I tried again.

When I started baking bread, I renewed the struggle for pizza. I quickly settled on using the bread dough for pizza. But I still had crustal difficulties. Try as I might, the crust always came out limp. We had an electric oven which warmed up slowly and the heat never seemed hot enough or dry enough. After numerous trials, I hit on the following strategy: First, I put the pizzas in a cold oven. That way, the heat came from below for most of the cooking, concentrating on the crust. Second, when the crust was cooked on the bottom, I slipped the pizza out of the pan and onto the oven rack for a few minutes to thoroughly stiffen and brown the crust.

We're up to 1982 now. I have a gas oven. I've been working on pizza for ten years.

The problem now was the topping. Everything I tried seemed to come out wet—except the pepperoni. My crust was crisp on the bottom but muddy on top.

Mushrooms were a special difficulty. I hated the idea of canned mushrooms, even though that's what I got in a pizza parlor. But fresh mushrooms produce puddles of water when they cook. I finally hit upon the strategy of cooking the mushrooms first.[3] Other vegetables also leaked. It took me some years to realize that I could slice the peppers more thinly and heap on fewer.

By the late 1980s I had a decent product. I could reliably make pizza, even for company. Confidence trumped humility. I made

---

[2] But not beyond the scope of the website. You'll find it there, together with a t-shirt inspired by the more general theory.

[3] Cook them in a flavorful fat. Beef fat is very good.

good pizzas and I developed successful combinations. I was content. I was sure. I stopped making progress.

But in 2002 something clicked. My pizzas were great (I thought). My crust was praised (true). But guests never cleaned their plates. They always left ends of crusts. Eventually, I doubted.

Doubt led me to wonder if I was using too much dough. Asking the question took a lot of humility. With a big wad of dough, I could show off. I could stretch, whirl, and toss the thing like a professional. With a smaller piece, I had to roll it out. But the results were a great improvement.

Pizza, basic, 16 inches

1/6 recipe bread dough (see chapter 2)
3 Tb tomato puree (not tomato paste!)
1/2 tsp oregano
1/4 lb mozzarella cheese
your topping, but not too much of it
2 Tb freshly grated Parmesan cheese

Grease the pan with butter.[4] Roll out the dough. This isn't easy. I roll it to a diameter of 13 to 14 inches, plop it into the pan, then stretch it gently up to the edge.

Smear the tomato puree on the surface of the dough and sprinkle with oregano. Top with the mozzarella, your topping, and sprinkle on the Parmesan.

Bake at 450° for 15–20 minutes. Check it occasionally. At the end, I remove it from the pan and cook it a minute or two on the oven rack.

All the usual items can be toppings:
* pepperoni—slice thinly so it comes out crisp
* mushrooms—cook them first
* onions—cook these first too, caramelizing them
* peppers, olives, marinated artichoke hearts, cooked hamburger, etc.

My pizza continues to evolve. I'm much lighter on the toppings now. This allows the crust to get cracker-crisp.

---

[4] Or not. I find now that grease is unnecessary. The cooked crust doesn't stick to the pan.

Grape Tomato Pizza

a cup or so of grape tomatoes
1/2 tsp chopped rosemary
2 to 3 oz good cheddar
optional: 1/4 cup pecans
olive oil
salt to taste

Split the grape tomatoes and distribute them on the pizza. Sprinkle with rosemary and cheddar. A "good" cheddar here is one with a lot of flavor, not necessarily very sharp. Vermont or New York cheddar is excellent. Sprinkle on the pecans if using.

Drizzle the olive oil. Salt the pizza. Bake.

Fennel Pizza

1 medium onion, very thinly sliced
1 cup very thinly sliced fennel bulb
3 Tb butter, or more
1/4 to 1/3 cup golden raisins, chopped
1 Tb cider vinegar
optional: chopped, lightly toasted hazelnuts
salt to taste

Saute the onion and fennel in the butter very slowly until the vegetables are dark and caramelizing. Add the raisins and vinegar and cook just a little more, then spread on the pizza. Sprinkle on the hazelnuts if opting.

Salt the pizza. Bake.

Reuben Pizza

Russian dressing
sauerkraut
slices of corned beef
Swiss cheese
freshly ground caraway seeds

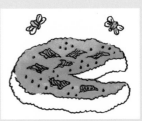

continued

---

continued

If possible, use a little rye flour in the dough.

Spread the Russian dressing on the pizza as you would tomato sauce. Top with the sauerkraut, the corned beef, the Swiss cheese, and then the caraway seeds.

Bake.

This last pizza is often the most appreciated.

Apple Pizza

apples, peeled, cored, and sliced
pinches of nutmeg or mace
good cheddar cheese
walnuts (not optional)
1 to 2 tsp sugar
1 to 2 Tb butter

Distribute the apple slices on the crust. Sprinkle with nutmeg or mace. Distribute the cheese. Distribute the walnuts. Sprinkle on the sugar.

Dot with butter.

Bake.

The lesson here is that you always have to wonder if your product/ idea/dish/theorem could be improved. Probably you already knew this. Probably you understand this better than the author, who needs to be reminded, from time to time, not to be so smug.

13

THE CLUELESS GEEK

Solving a challenging math problem is like baking a soufflé—you have to be prepared to risk failure if you want success.
—Keith Devlin (NPR's "Math Guy")

MY METHOD for solving problems could be described as:

1. Don't worry, just try something.
2. If that works, great, go on to step 3. If not, return to step 1.
3. Okay, now worry. You must have goofed. Find your mistake.

In short, I'm telling you to *mess up*. To most people that sounds wrong. It sounds like a bad strategy on a math test and a bad way to cook a dinner for ten. But in fact, a willingness to make mistakes is crucial to progress of any sort.

MESSING UP IN THE KITCHEN

Cooks don't make mistakes. They conduct experiments! To call it a mistake is an attitude problem. Don't change your procedure. Change your attitude.

Even an experiment that fails tells us something valuable. It took me five years of experiments to create the recipe for shamburgers. It took me twenty years of experiments to create the recipe for pizza (which I didn't give to you because after another ten years of experiments I found one I liked better).

In the case of pizza, I think I can give you a strong argument for mistakes. Pizza isn't simple. I don't think I've ever seen a pizza recipe advertised as "foolproof." You can read up on pizza but words can't completely prepare you. You have to jump in and mess up.

While trying to make pizza, I did use recipes, sometimes. The pizzas usually didn't turn out the way I (or the cookbook authors) wanted. But each time I learned more about dough, more about ingredients, more about oven thermodynamics, and a little more about what my family likes and will eat. I learned slowly. But I learned.

Currently, I am conducting experiments in ravioli, puff pastry, smoked ribs, and potstickers.

Noncooks often have the feeling that the ability to cook is innate. They say to themselves, "I don't know what to do here. If I had the genes to be a cook, I would know. I don't know, so cooking is beyond me."

Wrong. It takes time. It takes time for everyone. Put in the time (maybe a lot of it) and you'll succeed.

MESSING UP MATHEMATICALLY

When I started work on clueless sudoku (see chapter 3), I had no idea whether such puzzles were even possible. But I didn't wait for inspiration to strike, I jumped in and tried to make a puzzle. I had one failure after another. I learned a little from each.

When I started work on bouncing in a rectangle, I was totally mystified. But I played with rectangles, trying to see a pattern. One reason I was mystified was that instead of bouncing like this,

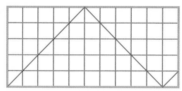

I was looking at the rectangle as a checkerboard,

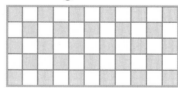

and I was bouncing like this:

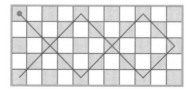

Instead of like this:

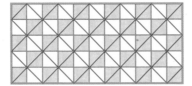

That checkerboard bouncing

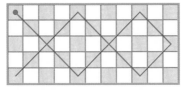

is really what happens in a smaller rectangle:

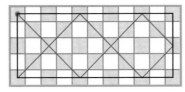

I got very confused. It was days before I realized what was going on. Did I waste my time?

Not at all! I couldn't tell you then and I can't tell you now exactly what I was learning. But in subtle, indescribable ways, my understanding was growing. I had to put in the time.

Nonmathematicians often have the feeling that the ability to do math is innate. They say to themselves, "I don't know what to do here. If I had the genes for this, I would know. I don't know, so math is beyond me."

Wrong. It takes time. It takes time for everyone. Put in the time (maybe a lot of it) and you'll succeed!

HOW MOZART MESSED UP

There is a wonderful anecdote about Mozart in *Mozartiana* by Joseph Solman.[1] Mozart was asked by a young man for tips on writing a symphony. Mozart said that symphonies were complicated and suggested starting with something easier.

---

[1] Walker, 2002.

"But Herr Mozart," the fellow answered, "you wrote symphonies when you were younger than I am now!"

Mozart responded, "I never asked how."

The difference between the young man and Mozart wasn't knowledge. It was *daring*. Mozart was unafraid to jump in and mess up. The young man was cautious. Because this young man lacked that essential daring, Mozart judged he had to start with something easier!

But enough messing up. Let's talk about *beauty*.

14

ELEGANT DISHES

Mathematicians sometimes use the word "elegant" to describe
the grace and felicity with which the elements of a mathematical proposition connect. It may seem far-fetched to borrow the
term and apply it to this most humble soup, but I believe it fits.
It is certainly not elegant in the sense that it is fancy. It is elegant
in the way the different properties of its meager ingredients are
explored, developed, and exquisitely related. The procedure is
simple, and nothing is wasted; there are no loose ends.
—*Marcella Hazan*

BEAUTY IN gastronomy is taste. Beauty in mathematics is more difficult to describe, but it's there. There is taste in mathematics. There
are aesthetics, there are concepts of beauty. The quintessential
mathematical aesthetic is elegance. Marcella Hazan, in the quotation above, has the sense of the word perfectly.[1] A few elements,
perfectly blended, nothing wasted, nothing extraneous.

CHENEY'S CARD TRICK

A lovely example of mathematical elegance can be found in a card
trick devised in the last century by William Fitch Cheney Jr, a mathematician at the University of Hartford. It is, first of all, a pleasingly
puzzling trick. Second, the mathematics behind it just manages to
make it work. There are just enough cards in a suit, just enough suits
in a deck.[2]

It takes a team to do the trick. Here's how it appears to an audience: One person, call her Deirdre, is out of the room. The other, call

---

[1] *More Classic Italian Cooking* (Knopf, 1978).

[2] A complete description of the trick, its history, and some intriguing variations appear in Colm Mulcahy, "Fitch Cheney's Five Card Trick and Generalizations," *Math Ho-
rizons*, February 2003, pp. 10–13.

him Harold, stands before the audience. Someone in the audience cuts the deck and shuffles it as often as desired, then deals Harold five cards.

Harold looks at the cards briefly and chooses one, which he gives back to the audience. He then puts the other cards in a pile on the table and leaves by one door as Deirdre enters by another.

Deirdre walks to the front of the room. She picks up the four cards left by Harold.

In a few seconds, she announces the name of the card that was left with the audience.

Amazing.

It works because Harold manages to signal to Deirdre, by the cards he leaves and their order, the identity of the fifth card. It is elegant mathematics.

Here's how it works: When Harold looks at the five cards, at least two of them must be in the same suit. In this case, there are two hearts. Harold chooses one of the hearts for the card to be returned to the audience. The other he puts as the first of the four cards he will leave in a pile. Thus, when Deirdre picks up the pile, she immediately knows the suit of the fifth card.

But which card in that suit?

Harold uses the other three cards to signal the denomination of the missing card. Think of the entire deck as being ordered as follows: all the clubs come first, then the diamonds, then the hearts, then the spades (it's alphabetical—c-d-h-s). In each suit, aces below twos, twos below threes, . . . , all the way up to kings. In other words:

$$A\clubsuit < 2\clubsuit < \ldots < A\blacklozenge < 2\blacklozenge < \ldots < A\heartsuit < 2\heartsuit < \ldots < A\spadesuit < 2\spadesuit < \ldots < K\spadesuit.$$

In the example above, we have three cards to arrange, 7$\clubsuit$, 3$\blacklozenge$, and 8$\spadesuit$ (one of the hearts is with the audience and we put the other heart first in the pile). These cards can be ordered in exactly six different ways:

7$\clubsuit$	3$\blacklozenge$	8$\spadesuit$
7$\clubsuit$	8$\spadesuit$	3$\blacklozenge$
3$\blacklozenge$	7$\clubsuit$	8$\spadesuit$
3$\blacklozenge$	8$\spadesuit$	7$\clubsuit$
8$\spadesuit$	7$\clubsuit$	3$\blacklozenge$
8$\spadesuit$	3$\blacklozenge$	7$\clubsuit$

If we think of the lowest card as L (in this case 7$\clubsuit$), the card in the middle as M (3$\blacklozenge$ here), and the highest card as H (8$\spadesuit$), then the orderings are

1	–	L	M	H
2	–	L	H	M
3	–	M	L	H
4	–	M	H	L
5	–	H	L	M
6	–	H	M	L

Harold can use the ordering to signal to Deirdre one of six possible numbers. Deirdre can take the first or bottom card (the one that signals the suit) and add the number of the ordering to it. In that way, she can signal one of six different cards. In our example, the cards Harold left are

$$2\heartsuit \quad 7\clubsuit \quad 8\spadesuit \quad 3\blacklozenge.$$

From the 2$\heartsuit$ Deirdre first knows the missing card is a heart. Next she looks at the three remaining cards. Their order is M L H. That's order number 3, so Deirdre adds 3 to 2 and announces: "The five of hearts!"

Suppose the cards that Deirdre sees are

$$10\diamond \quad Q\spadesuit \quad 5\clubsuit \quad 7\spadesuit.$$

Deirdre knows the missing card is a diamond. The order of the last three cards is H L M, order number 5. Then Deirdre should add 5 to 10, getting 15. This isn't a card, but if you think of the cards as being in a circle,

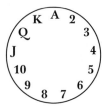

then adding 5 to 10 means starting at 10 on the card circle and moving clockwise five cards, arriving at 2.

Deirdre would announce "The two of diamonds!"

There's one problem, though. Harold can signal adding 1, 2, 3, 4, 5, or 6 to the first card, but there are actually 12 possibilities for the fifth card. Suppose, for example, that the five cards Harold picks up are:

$$8\clubsuit \quad 4\heartsuit \quad 3\diamond \quad 7\spadesuit \quad J\heartsuit.$$

The two cards of the same suit are the 4 and the jack of hearts. If Harold gives the audience $J\heartsuit$, then he can't signal to Deirdre $J\heartsuit$, because the distance from 4 to J on the card circle is 7 (bigger than 6).

The solution is that Harold *doesn't* give the audience the jack; he gives the audience the 4, because you can get from J to 4 on the circle by adding 6.

So Harold leaves

J♥ 7♠ 3♦ 8♣,

since 7♠ 3♦ 8♣ is H M L, the sixth order, and J + 6 = 4 on the card circle.

Do you see how elegant this is? Out of five cards there will be two of the same suit. *With one more suit this wouldn't work.*

Of those two cards, one of them will be six or fewer ahead of the other on the card circle. *One more card in a suit and this wouldn't work.*

Harold chooses that card to hand back to the audience. Now there are just six possibilities for the missing card, but Harold can signal which one by arranging the remaining three cards because there are six possible arrangements. *One fewer arrangement and this wouldn't work!*

The line is thin between elegance and magic. It's most appropriate that the elegant mathematics here enables an elegant magical effect.

FRED'S GRILLED CHEESE SANDWICH

The quotation at the beginning of this chapter is about Hazan's recipe for broccoli soup. She writes further,

> The blanched broccoli is sautéed with garlic in olive oil. The florets are kept aside, but the stalks are puréed together with their oil, and added to the broth and the egg barley. The density of the soup is thinned out with some of the water in which the broccoli was blanched. When the pasta is done, the florets are dropped in, and the soup is done. The tenderness

of the florets, the firmness of the pasta, the savoriness of the good broth enriched with oil, and the faintly garlicky puréed stalks, all fall nimbly into place. If there is a better word than "elegant" for how it is done, I can't think of it.

Another recipe I would term elegant is my son's recipe for a grilled cheese sandwich. Before I explain his method, let me list the wonderful properties of his sandwich:

- A good grilled cheese sandwich should be crunchy. This is crunchy.
- But grilling in oil makes a sandwich greasy. This isn't greasy.
- Sometimes a grilled cheese sandwich is so crunchy it lacerates your gums as you eat it. That doesn't happen with this sandwich, because it's soft.

Soft? You just said the sandwich was crunchy!

- It's crunchy *and* soft. It's amazing.
- Bread is tastier when toasted. So is cheese, but in a standard grilled cheese sandwich, the cheese isn't toasted. But it *is* toasted in this sandwich.

Okay, here's the magic:

Fred's Inside-grilled Cheese Sandwich
(For two sandwiches)

4 slices bread
cheese for two sandwiches
toaster oven

Place the bread and cheese in the toaster oven like this (two slices of bread in each pile, one layer of cheese):

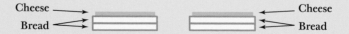

Then turn the toaster oven on to toast. Toast until either the bread is as dark as you want or the cheese is as toasty as you want, whichever comes first. The outside will be toasted but the inside will be soft.

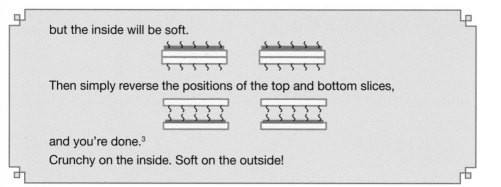

Then simply reverse the positions of the top and bottom slices,

and you're done.[3]
Crunchy on the inside. Soft on the outside!

Mathematicians think of elegance as being a special mathematical sort of beauty. But it's universal. A dish can be elegant in the mathematical sense too.

But enough of beauty! It's time to be useful.

---

[3] You can make these sandwiches in a broiler too. Place the four slices of bread, two with cheese on top, on a cookie sheet under the broiler.

15

FOOD FOR THE MASSES

THERE'S A practical side to cooking. The original purpose of cooking was to give sustenance, to provide healthy, sustaining meals.

And there's a practical side to mathematics. The original purpose of mathematics (arithmetic and geometry) was to solve important problems in society, in agriculture, in religion.

It's clear, then, that mathematics and gastronomy are similar in that they are of material importance to the world. It's so clear, in fact, that I won't mention it again.

This book is devoted to making clear that the immaterial aspects of mathematics and gastronomy—

- what attracts us to them
- how we judge them
- how we pursue them
- where we get our ideas

and a host of other issues—are also remarkably similar.

That's enough of being useful! On to the next chapter.

16

GOOD FOOD FOR THE MASSES

BUT I do want to acknowledge that even if we focus on pleasure, we can't ignore the practical side of cooking and math, for usefulness itself is an attraction, as much as simplicity, complexity, or elegance.[1] It's a great pleasure when one's work in the abstract world of mathematics has meaning in the world outside, most especially when it contributes to human happiness. And it's a pleasure when a kitchen discovery is important to others, if, say, it improves health by creating delicious and healthy foods, or it alleviates hunger by inventing more efficient uses of produce.

One is tempted to say that all cooking is useful. That hardly seems true of mathematics. But then I will quote my colleague Jim Callahan:

> There are two kinds of mathematics: applied mathematics and mathematics that is not yet applied.

> There are examples of mathematical theorems hundreds, even thousands of years old, which have become useful only recently. While it's hard to imagine practical applications for many, if not most, mathematical theorems, it is impossible to rule such applications out.

But enough of that. It's time to play!

---

[1] Or, coming up, playfulness and weirdness.

17

JUST FOR FUN

PLAYFULNESS IS abundant in the preparation of food. It's also a mathematical aesthetic.

THE PLAYFUL MATHEMATICIAN

The reader may feel that most of the mathematics in this book is playful: puzzles, games, doodles, and card tricks. That may reflect a bias in me. Like any aesthetic, it is shared by some but not all.

A lovely example of mathematical playfulness is the following theorem discovered and proved by Bancroft Brown.[1]

> **Theorem:** *The thirteenth of the month falls on a Friday more often than any other day.*

A strange statement! You would think that the thirteenth would fall more or less equally on all the days of the week!

The statement is also playful. It's not about polyhedra, the fourth dimension, general relativity, or infinite series. The insignificance of the statement is part of its charm.

The statement is useful, too, if you happen to be superstitious.

And finally, the proof is cute, cute with a touch of madness. Here it is:

> **Proof:** The number of days in a year varies according to the following rules:

---

[1] I heard this at a lecture Brown gave at Dartmouth as professor emeritus. Brown was a geometer at a time when geometry was not in fashion. He wanted to tell us why he had chosen this field that others had left. The theme was essentially "We had a lot of fun." The theorem here is not geometrical, but it is fun. Geometry, by the way, is back in fashion.

1. There are 365 days in a year *except*—
2. Every fourth year there are 366 days (leap year) *except*—
3. Every one hundredth year there are 365 days *except*—
4. Every four hundredth year there are 366 days.

(Example: 1885, 1886, 1887 each had 365 days [rule 1], but 1888, 1892, and 1896 were leap years [rule 2], but 1900 was not a leap year [rule 3]; 2000, however, was a leap year [rule 4].)

From these rules, we can see that there are a fixed number of days in any 400-year period. We can calculate the number of days in 400 years by using the four rules:

1. 365 days times 400 years equals 146,000 days.
2. 146,000 days plus 100 leap years equals 146,100 days.
3. 146,100 days minus 4 leap years equals 146,096 days.
4. 146,096 days plus 1 leap year equals 146,097 days.

Surprisingly, this number is divisible by 7:

$$146,097 = 7 \times 20,871.$$

That means, for example, that since April 3, 2003, fell on a Thursday, then April 3, 1603, also fell on a Thursday and April 3, 2403, will fall on a Thursday too. It also means that the number of Monday the thirteenths during the period from 1600 through 1999 is the same as the number of Monday the thirteenths during the period from 2000 through 2399 is the same as the number of Monday the thirteenths during the period from 2400 through 2799, and so on.

Thus, to show that the thirteenth falls on a Friday more often than any other day, all we have to do is show that this is true in the period from 1600 through 1999 (or in any other 400-year period).

That's the cute part of the proof. The mad part of the proof is that Brown, having gotten this far, spent hours (days?) computing the number of Monday the thirteenths, Tuesday the thirteenths, Wednesday the thirteenths, etc., for the period 1600 through 1999. He did this before computers were available to do the work. When he was done, he found that the thirteenth fell on a Friday more often than any other day during that particular 400-year period. Hence, it falls on a Friday more often than any other day in *any* 400-year period!

THE PLAYFUL CHEF

Play is appreciated in food, of course. Have you ever been served a fish with its tail in its mouth? Have you ever been served a cake with prizes inside? What about fortune cookies?

A few years ago I had the idea of baking my own version of the Hostess cream-filled cupcake. I succeeded; it was playful. Later I learned that others were doing that too. So instead of offering that recipe, I offer something no one else (I think) has imagined.

For years I've thought that mangoes and cashews were a nice pair. I've made mango tarts with a cashew crust. That's good, but the dessert I'll show you in a moment is better. The idea came to me when I thought how much a scoop from a mango looks like an egg yolk. That led me to the idea of dessert that looks like eggs Benedict—but isn't.

The dish consists of a nut meringue (looking like an English muffin), a panna cotta (looking like an egg white), mango (looking like an egg yolk), a *crème anglaise* (looking like Hollandaise sauce), topped with some poppy seeds (looking like ground pepper).

Ex-Benedict

For the panna cotta:
    1 envelope unflavored gelatin
    3 Tb cold water
    1 1/4 cups heavy cream
    1 1/4 cups milk
    1/3 cup sugar
    1 vanilla bean or 1/2 tsp vanilla extract
    six 4- to 6-oz custard cups, lightly oiled

Sprinkle the gelatin over the water and leave it for five minutes.

Place in a saucepan the cream, milk, and sugar. If using a vanilla bean, split it lengthwise, remove the seeds, and place it in the saucepan. Bring the ingredients to a boil, then turn off the heat.

If you see a vanilla bean, remove it; if you don't, add the vanilla extract.

Add the gelatin, stirring until it dissolves. Let this cool a bit, stirring as it starts to thicken. Pour the mixture into the custard cups. Make an air-tight cover by placing plastic wrap over the panna cotta, with the plastic resting on the surface of the custard.

For the biscuits:
   3 egg whites
   3/8 cup sugar
   1/2 cup + 1 Tb raw, unsalted cashews, finely ground in a coffee grinder
   1 large tuna can, clean, top and bottom removed

Cover a baking sheet with buttered foil. Beat the egg whites until soft peaks form. Beat in the sugar a tablespoon or two at a time. Fold in the ground cashews. One at a time, form six biscuits on the buttered foil using the tuna can to shape them.

Bake at 250° until golden. This may take an hour or more. Or less! Check on them every 10 minutes or so.

Let the biscuits cool. Then gently peel off the foil.

For the crème anglaise:
   3 egg yolks
   1/2 cup sugar
   3/4 cup hot milk
   1/2 tsp cornstarch
   1/2 tsp vanilla

Beat the yolks, beat in the sugar, beat in the cornstarch, beat in the hot milk *gradually*. Stir in the vanilla. Cook, beating, over hot water until it thickens.

For the assembly:
   3 Ataulfo (also called Champagne) mangoes
   poppy seeds

Place the biscuits on plates. Loosen the panna cottas from the custard cups by running a knife around the edge. Carefully invert the cups and nudge the panna cottas out with the knife. Place them on top of the biscuits. Slice each mango twice, once just above the central flat seed and once just below the seed. From each of the six "cheeks" scoop out a round of mango. Place the scoops on the panna cottas, flat side down. Spoon on crème anglaise. Sprinkle a pinch of poppy seeds.

This dessert makes a pretty good impression. It's not hard to see the resemblance to eggs Benedict. But the best reaction I've had is when I told my guests, before I brought it out, that the dessert was *not* eggs Benedict. When the dish was put before them, their faces fell. Because I told them it wasn't eggs Benedict, the idea of eggs Benedict was in their heads. They had just eaten a large meal and they didn't want to eat an eggs Benedict. Worse, the dessert is big, like eggs Benedict on steroids.

Of course, they hugely enjoyed it in the end. Despite its appearance, it's a fairly light dessert.

My message here is that it's good to have fun. Most people know this.

18

JUST TO BE WEIRD

WE'VE DISCUSSED several aesthetics in cooking and mathematics:

Elegance
Simplicity
Complexity
Usefulness
Playfulness

There's (at least) one more: Weirdness.

THE WEIRD CHEF

Many things are imagined and some of them are cooked, served, and eaten. I don't mean the bizarre (but traditional) dishes of cultures in far-off places. I mean invented dishes. Dishes dreamed up to shock and delight.

There's chocolate-covered bacon, tobacco rhubarb, garlic ice cream, chicken with vanilla, coffee-rubbed cheeseburgers.

There is a recipe that calls for burning toilet paper in a can of tuna (to smoke it).[1]

These recipes are intended to be enjoyed, that is, they are meant to be delicious. But in addition, they give the pleasure of the unusual, the weird.

To these, I have my own contribution. You've heard of red pizza and white pizza. I suppose a pesto pizza is a green pizza. But now there is a . . .

---

[1] I've had this, it's pretty good. See http://brokeassgourmet.com/articles/smoked-tuna-salad.

Blue Pizza

1/6 recipe bread dough (a third of a loaf) (p. 15; see also p. 96)
1/3 cup blueberries
2 oz blue cheese
2 Tb sugar

If possible, use wild blueberries, the small, low-bush variety. These have less juice (this pizza can be dangerously soggy). The cheese should be mild but definitely blue.

Distribute the blueberries on the crust. Crumble the cheese and distribute that too. Sprinkle on the sugar. Bake at 450° for 15–20 minutes.

I've also invented a green dish that's unusual if not conclusively weird. It features a lot of green tastes.

A Green Amuse Bouche

1/3 cup fresh oregano, loosely packed
    or 1/2 cup fresh basil, loosely packed
    or a loosely packed combination of
        basil and oregano (the theme here
        is chlorophyll)
1 tsp (green) Chartreuse
juice of half a lime
1 tsp especially fruity olive oil
1/4 tsp sea salt
2 tsp pink peppercorns (I know. Not
    green. But good.)
2 avocados, slightly underripe

CHARTREUSE

Chop the herbs well and pound with the salt and pink peppercorns. Add to the liquid ingredients.

Just before serving, dice the avocado and add the mixed ingredients.

Serve on tiny plates or cups.

Having invented green and blue dishes, I want to expand to red, yellow, and brown (but remaining weird). No ideas, yet.

WEIRD MATH

We do like mathematics in which all the pieces fit together beautifully. "Elegant" is high praise for a theorem or a proof (see chapter 15).

But mathematicians also enjoy surprises, results that defy expectations. "Bizarre!" is often used to express pleasure and admiration for a work of mathematics.

Examples of the bizarre abound in mathematics. To the ancients, the discovery of irrational numbers was famously bizarre.

The discovery, in the eighteenth century, that a curve, which we think of as one-dimensional (it has length but no thickness), can be so wiggly as to be two-dimensional and fill up a square, is another example. Of course, what is weird initially becomes, in time, part of our deeper understanding. Whenever we come across something strange, we know we're going to learn something important.

Here is something bizarre that I came across fairly recently. You recall how we charted the path of a bouncing point in a rectangle, and we saw that the length of the path is simultaneously a multiple of the width and the length (chapter 8).

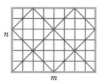

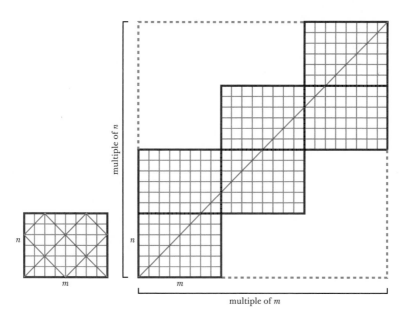

That means that if we try a rectangle where there is no common multiple, say $n = 6$ and $m = \sqrt{70}$, then the path

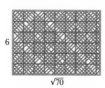

will go on forever.

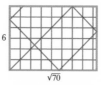

This isn't, by the way, what we meant by a curve that fills a square. The path in the rectangle goes on forever, but it misses many points. A square-filling curve doesn't miss any points. It also doesn't cross itself. It's bizarre.

Later, we looked at boxes and we followed the path of a point moving around the box.

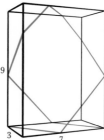

It seems reasonable that if one or more of the dimensions were irrational, then the path would go forever. So if we take the wildest box imaginable,

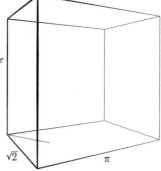

($\sqrt{2}$ = 1.4142136 . . . , π = 3.1415926 . . . , e = 2.718281 . . .), then we expect that the path will go on forever. But it doesn't. It actually returns to the starting point!

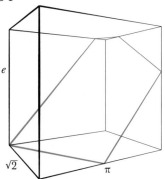

And we can learn something from this.

We understood the rectangle better when we drew reflections of it and followed the path in those reflections. We can do almost the same thing with the box. What we do is (sort of) unfold it. Here's the base and one side folded down:

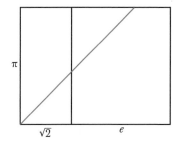

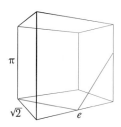

And now we add another side:

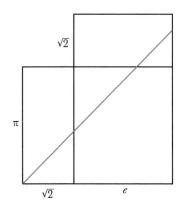

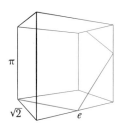

And then the top:

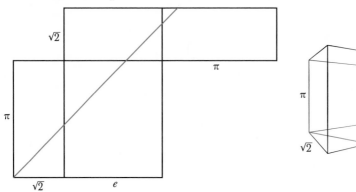

Then finally another side:

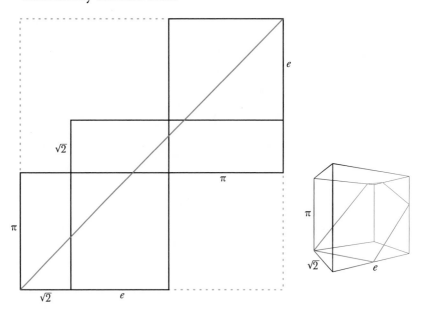

And it all makes sense because the diagram sits inside a square with side equal to $\sqrt{2} + e + \pi$.

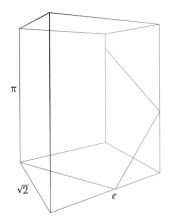

By the way, if you rotate the box so you're starting on a different face,

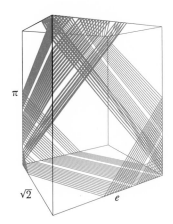

then the path does go forever.

But not everywhere. It lives on just a part of the box.

Weird!

Weird and cool.

19

CELEBRITY CHEFS

ARE THERE celebrity mathematicians? There are a few. The hottest might be Vi Hart.[1] But there's also Danica McKellar and, for the *Weekend Edition* crowd, the "math guy," Keith Devlin.[2]

But there's nothing in the math world remotely like the cooking channel. Mathematicians aren't often on camera. You don't see them on talk shows. They have little Klout.

That said, two figures of the last century, one in cooking and one in mathematics, seem roughly comparable. And they were, both of them, transformational figures.

JULIA

Julia Child (1912–2004) had a major impact on the way we cook and the way we eat. She enabled a revolution.

First, let me sketch a grossly simple picture of the gastronomic state of the nation in 1960. Americans ate from a narrow range of foods and flavors. Diners and cooks were unadventurous. Ethnic restaurants tailored dishes to suspicious and xenophobic palates. Home cooks strove only to prepare the standard dishes well, and with little fuss. Cooking was a chore that modern devices were working to reduce.

"French" cookbooks offered recipes that simplified dishes. Ingredients were substituted so that everything could be found in a neighborhood grocery store. Procedures were simplified so that meals could be prepared in an hour. The result was that few people

---

[1] www.youtube.com/user/Vihart. By the way, Vi Hart has started creating mathematical food. Her flex Mex is spectacular.

[2] http://www.danicamckellar.com/; http://www.stanford.edu/kdevlin/.

knew the pleasures of French cuisine. Still fewer had any inkling that they could produce them.

Julia's first book, *Mastering the Art of French Cooking*, with co-authors Simone Beck and Louisette Bertolle, expanded American appetites and ambitions.[3] It was written so clearly that novices could follow the instructions and reliably produce classic dishes. Americans were introduced to a new world, a world with subtleties of taste they had never imagined. Most important, they learned that they could, with careful work, prepare these dishes themselves.

As an example, let's take *coq au vin*, one of the few French dishes that Americans were familiar with in 1960. The recipe in *Joy of Cooking* is typical. The ingredients are mostly there, but without much attention. Any sort of onion is allowed. You are told you can use either red wine or sherry. The brandy is optional. Then all the ingredients are cooked together in one pot. Preparation time is about an hour and a half. The Chamberlains' version in *The Flavor of France* is similarly loose. The recipe in *Amy Vanderbilt's Complete Cookbook* uses canned mushrooms and gets the cooking time down to an hour.

Julia's recipe, by contrast, takes all day, if you count preparing a brown stock for the onions. The bacon alone requires four stages. It is cut into rectangles, simmered in water, rinsed and dried, and then saut´eed in butter, all before it is added to the chicken. The small white onions are browned in butter and then carefully braised in a brown stock with herbs. The mushrooms are separately browned in butter and oil. Both are added to the dish only on the serving plate. It's a lot of work. But the result is heavenly.

Julia's television program accelerated the change. Having aroused the nation's taste and desire, she now persuaded Americans in great numbers to tackle the classics of French cuisine. Her messages were:

- Anybody (including you) can do it.
- You can make mistakes, even big mistakes.
- Food can be really, really good.
- Cooking is a lot of fun.

Julia led a revolution. Ironically, the fraction of people cooking today is significantly smaller than in 1960. Many Americans get their

[3] Volume 1 (Knopf, 1961).

food regularly from restaurants and take-outs. Supermarkets have huge selections of hot and cold meals ready to go.

And yet, I suspect, more of us cook to please ourselves today (as opposed to cooking to feed ourselves). That is, in part, Julia's legacy:

> Learn how to cook—try new recipes, learn from your mistakes, be fearless, and above all have fun!   —Julia Child

MARTIN

Martin Gardner (1914–2010) had a major impact on how we view mathematics, what mathematics we do, and even who does mathematics. He facilitated a quiet revolution.

First, let me sketch a grossly simple picture of mathematical practice and its public face in 1960. To the average American, mathematics was just the stuff taught in high school and college. It was a body of knowledge, useful and unattractive, ancient and unchanging.

Mathematicians themselves saw a more exciting field, but for the most part they worked in a few well-defined areas: algebra, analysis, topology, number theory, and geometry. Mathematical research was done by professional mathematicians, men (usually) with doctorates in mathematics.

From 1956 to 1986, Gardner wrote a column in *Scientific American* that slowly and subtly changed that. He wrote about new mathematics. The topics didn't fit in the standard categories. He wrote about games, puzzles, magic tricks, and doodles. He published problems that no one had solved. His readers wrote back. He reported on their difficulties and their successes.

As his readership grew, the perception of mathematics, both inside the mathematical community and out, began to change. As he reported progress in magic squares, plane tiling, secret codes, and knot theory, these areas acquired new status. When he discussed takeaway games, magic tricks, philosophical and logical paradoxes, paper-folding and paper-cutting, what had been considered periferal acquired weight and significance.

The greatest impact, however, of Gardner's writing was the revolutionary idea that mathematical research was a community affair. He invited his readers to work on unsolved problems and they responded. He was a clearinghouse for compelling mathematical problems, small and large. Amateurs joined in. His column gave readers a glimpse of a wonderful world where mind-twisting problems were posed, where ideas were freely exchanged, and where results were excitedly announced. You wanted to join that world. You could join that world.

Martin Gardner did much more. He wrote more than 100 books on mathematics, philosophy, religion, and science. He annotated works of literature. He wrote short stories and novels. In a life spent writing about mathematics, his messages were:

- Anybody (including you) can do it.
- There is a lot of cool mathematics.
- Cool mathematics is good mathematics.
- Mathematical research is serious play.

To his mathematical fans, he was simply Martin. Years after his death, he is remembered in a biennial conference, the Gathering for Gardner, celebrating his many interests, magic, puzzles, mathematics, philosophy, and scientific fallacies.

Martin led a revolution. Ironically, the mathematical preparedness of Americans today is widely seen as lower than forty years ago. High school graduates depend on calculators. More than half of the mathematics doctorates granted in the United States today are earned by foreign scholars.

And yet there is more interest and enthusiasm now for mathematics, and more participation. Fractals, chaos, and RSA coding have entered the vocabulary. Millions watched *NUMB3RS*. Millions solve sudoku puzzles. While hardly anyone wrote about mathematics forty years ago, there is a strong market today for books on mathematics. That also is Martin's legacy.[4]

Martin has turned thousands of children into mathematicians, and thousands of mathematicians into children.[5]

---

[4] I'm not the only one to have noticed the parallels between Martin and Julia; Colm Mulcahy has written and spoken on this. See, for example, "Food for Thought: Savory Treats for the Mind from the Julia Child of Mathematics and Rationality," Huffpost, October 10, 2012.

[5] Persi Diaconis, mathematician and magician.

20

ECONOMY

MATHEMATICS AND gastronomy share an appreciation of economy, of doing things with the least effort.

ECONOMY IN THE KITCHEN

I did a little searching around and found:

> Six cookbooks specializing in recipes using six ingredients or fewer.
>
> Three cookbooks specializing in recipes using five ingredients or fewer.
>
> Four cookbooks specializing in recipes using four ingredients or fewer.
>
> Fourteen cookbooks specializing in recipes using three ingredients or fewer.
>
> One cookbook with recipes that use only two ingredients.

The chief attraction of such books is the promise of quick, easy-to-prepare recipes. Beyond convenience, though, there is an appreciation of economy. There is a directness, a neatness, an honesty in a dish composed of just a few essential components.

Most of the books I saw offer uncomplicated, quick recipes, but those by Rozanne Gold are different. She argues that many dishes are improved by reducing the number of ingredients. In her *Recipes 1-2-3: Fabulous Food Using Only 3 Ingredients*, she writes:

> So when we confront some overwrought dishes that today's chefs bring to the table, we often are not experiencing the layers of flavor we'd like to believe, but rather a muddling or masking of basic flavor components.[1]

[1] Penguin Books, 1996, p. 1.

Gold's recipes are not necessarily simple. In her most impressive dishes, nuance and complexity are extracted from simple materials with time and art.

A lovely example is her "Oven-Roasted Asparagus, Fried Capers." Medium-size asparagus spears are roasted at a very high temperature, 500°, for 8 minutes, drizzled with olive oil, sprinkled with salt, and served with large capers fried in olive oil. The unusual preparation of both the asparagus and the capers intensifies their flavors.

Salt, pepper, and water don't count as ingredients. That seems fair, especially for the salt and water. Pepper adds its own flavor, though. For that matter, what is an "ingredient"? Gold includes here some fairly complex concoctions: mayonnaise, pesto, Chinese five spice powder, veal sausage, cheese raviolini, chicken broth, and so on.

For the simplicity of its components (just four), I have long admired the Philippine dessert Sans Rival. It's a marriage of Spanish and Philippine cuisines, an Asian dacquoise. It's a heavenly dessert which should be better known.

Sans Rival

For the layers:
- 8 egg whites
- 1 cup sugar plus
- 1 1/2 cups finely ground cashew nuts (use a coffee grinder)
- aluminum foil
- butter for buttering

Cover three baking sheets with buttered foil. Beat the egg whites until standing in soft peaks. Beat in the sugar a tablespoon or two at a time. Fold in the ground cashews. Spread the mixture on the three buttered foils. When done, these layers will be stacked on top of each other so they should be close to the same size and shape. I just made some layers. I measured them at 13 × 9 inches each. They should be of uniform thickness so they will brown evenly (and all parts will be crisp).

Bake at 250° until golden all over. That may take as long as 1 1/2 hours. Let cool. Then gently peel the foil from the layers.[2]

continued

[2] Note: it's impossible to peel off the foil when the layers are hot.

continued

If a layer breaks, that's fine; the best layer can go on top.

For the buttercream:

8 egg yolks

2/3 cup sugar

2 sticks unsalted butter

a double boiler

a stand mixer

Beat the egg yolks.

Put the sugar in the top half of a double boiler with 3 tablespoons of water. Heat the water and sugar to 238° (this can be done over direct heat). Let it cool a minute or two. Then pour the syrup in a thin stream over the yolks while beating them vigorously. Place the top half of the double boiler on the bottom half of the double boiler with 1/2 inch water in the bottom. Cook the sugar–yolk mixture, stirring with a wooden spoon, until the mixture thickens. Traditionally, one stops when a spoonful dribbled over the surface forms a "slowly dissolving ribbon." That's tricky. If you cook the yolks too long, they harden—very bad. I usually turn off the heat when I see a quickly dissolving ribbon.

Cool the mixture over cold water, stirring occasionally, until it is just a little warm. Place it in the bowl of stand mixer and beat in the cold butter one tablespoon at a time.

Finally, spread the buttercream on the three layers and stack them. Save a little more for the top layer. Don't save any for the sides (which are difficult to cover).

Refrigerate the Sans Rival.

Serve it in small portions—it's rich.[3]

This is best the day it is made. In time, the layers lose their crispness. The combination of crisp and dry with fat and creamy is part of the wonder of this torte.

Some recipes call for corn syrup in the buttercream. Some flavor the buttercream with rum. The rum, I think, only interferes with the purity of the rest of the ingredients.

[3] But offer second helpings. Have one yourself.

I admire Gold's creative and varied cooking techniques—roasting beets, burning oranges, and smoking bay leaves. I also see in her, when she discusses making her own raisins, something of the madness that occasionally overtakes me.

A few of her recipes are reductions of classic dishes, which inspires the following puzzle.

PUZZLE

Three of the recipes in *Recipes 1-2-3* are three-ingredient versions of old standards. I'll list the ingredients of the standard recipes and invite you to guess the three ingredients of Gold's recipes. Don't forget—salt, water, and pepper don't count.

1. **Borscht**, a recipe of Mrs. Harold Eisenberg in *Russian Cookbook for American Homes*, Gaynor Maddux editor, 1942:

    Soup meat, soup greens, onions, beets, vinegar, butter, flour, tomatoes, and sour cream.

2. **Spoonbread**, from *Joy of Cooking* by Irma Rombauer and Marion Rombauer Becker, 1971:

    Corn meal, flour, sugar, baking powder, egg, milk, and butter.

3. **Osso buco**, from *The Classic Italian Cookbook* by Marcella Hazan, 1973:

    Onion, carrot, celery, butter, garlic, lemon, oil, veal, flour, white wine, meat broth, tomatoes, thyme, basil, bay leaves, and parsley.

The answers are at the end of this chapter.

MATHEMATICAL ECONOMY

Economy is a quintessential mathematical virtue. It's an aesthetic that mathematics shares with poetry. The greatest works of mathematics move from modest assumptions, briefly stated, to the most profound conclusions.

In the fourth century B.C., Euclid showed that all the mathematical knowledge of his world followed from just five postulates. That is brevity, certainly. But as I mentioned earlier, mathematicians tried, for two thousand years, to reduce the five postulates to four.

One branch of mathematical logic—called "reverse mathematics"—crystallizes the economical impulse. It attempts to discover, for certain theorems, exactly what axioms, postulates, or assumptions are necessary to prove the theorems.

This aesthetic can be seen clearly in the sphere of mathematical games. The assumptions of a game are the rules. A beautiful game is one where the rules are simple but the strategy is complex.

Consider the games of Chess and Go. Both are incredibly rich with long histories of analysis and study. Both are "good" in the sense described here—their rules are short and the consequences of those rules are long. But between the two games, Go is significantly "better." The rules of Go are simpler than Chess and simpler to state. All the pieces of Go are identical. The great complexity of Go is achieved with humbler material.

An equally economical game is *Phutball*, or *Philosopher's Football*,[4] invented by mathematician John Conway. It's played on a 19 × 15 grid. The rules are beautiful and simple, yet the game is immensely complex.

Players sit at opposite ends of the board. The game starts with a single black stone in the center. This is the "football."

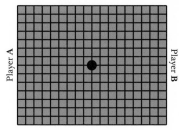

All the other pieces in the game are white and are placed on the board by the players.

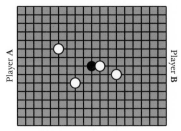

A turn consists of either

---

[4] Think of "football" as most of the world thinks of it, the game known as "soccer" in the United States.

- placing a white stone at an unoccupied intersection, or
- "jumping" white stones with the football.

The game is won when the football is jumped onto or over a player's goal line (the edge of the board away from the player).

You may jump a line of white stones in any of the eight directions, for example,

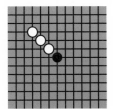 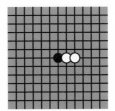 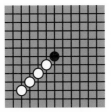

The stones that are jumped are immediately removed.
You may make multiple jumps in a single turn.

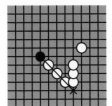 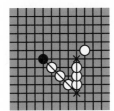 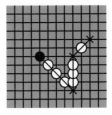

But no stone can be jumped twice—this move, for example, wouldn't be legal:

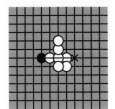 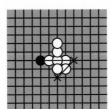 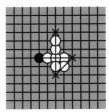

You may touch your own goal line during a move as long as you don't end your turn on it.

That's the game: Add a stone or jump the phutball.

Phutball is fun to play but devilishly hard to analyze. The game has recently been proved to be significantly more complex than either Chess or Go.[5] Play it yourself!

---

[5] See "Phutball Endgames Are Hard" by Erik D. Demaine, Martin L. Demaine and David Eppstein, in *More Games of No Chance*, Richard J. Nowakowski, editor (Cambridge University Press, 2002).

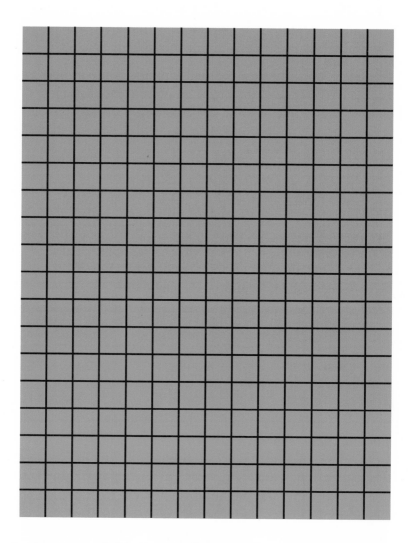

In the sense of this chapter, "economy" isn't really an aesthetic. It's a virtue.

The next chapter will continue a focus on virtue (balancing the previous focus on sin).

Answers to the puzzle:

1. Borscht: beets, squash, and yogurt
2. Spoonbread: cornmeal, eggs, and buttermilk
3. Osso buco: veal, tomatoes, and olives

21

ETHICS

SOME PEOPLE become vegetarian for health reasons. We've dealt with health already (chapter 15).

Some people become vegetarian because they dislike the taste of meat. We've dealt with matters of taste extensively (chapters 3, 4, 17, etc.).

Some people become vegetarian because they are worried about the effect of a carnivorous culture on the planet. We've dealt with the environment already (chapter 11).

But most people become vegetarian because they feel that the consumption of meat is morally wrong. Is there anything similar in mathematics? Are there ethical questions in mathematics?

ETHICS IN MATHEMATICS

Over the ages many have taken philosophical positions on mathematics that have a distinctly moral edge. Some have argued about what should be accepted as "number." Others have argued about what should be accepted as "proof."

Today, the mathematicians with the most well-developed ethical system are the constructivists. In their work, constructivists avoid using any "nonconstructive" method. A mathematical proof is nonconstructive if it uses an object without providing an explicit construction of the object.

Suppose, for example, I define a number n as follows:

$$n = \begin{cases} 0 \text{ if it rains in Pittsburg on April 5, 2371.} \\ 1 \text{ if it doesn't rain in Pittsburg on April 5, 2371.} \end{cases}$$

A constructivist would say that n has not been defined since we have no way, at present, of saying exactly what n is. It's not (at the present time) constructible.

But can't we say that whatever n is, it's less than 2? After all, n is either 0 or 1.

A constructivist would say no, we can't say n is less than 2. A constructivist might explain, "If you can prove to me that it will rain in Pittsburg on April 5, 2371, then we will know that $n = 0 < 2$. If you can prove to me that it won't rain in Pittsburg on April 5, 2371, then we will know $n = 1 < 2$. Otherwise, we know nothing."

The chief problem with constructivism (if you're not a constructivist) is that it's limiting. Constructivists can't enjoy some really lovely bits of mathematics. I'll tell you about one.

First, I ask you to recall that $\sqrt{2}$ is an irrational number, that is, it can't be expressed as n/m with n and m whole numbers.[1] Now, since we can write

$$\sqrt{2} = 2^{1/2},$$

this means we have an expression of the form

$$a^b$$

with both numbers, $a = 2$ and $b = 1/2$, rational but a^b irrational. The following theorem states that the reverse is also possible.

> **Theorem:** There exist irrational numbers c and d, not necessarily different, such that c^d is rational.

Here's the proof—does it convince *you*?

Proof: Consider

$$\sqrt{2}^{\sqrt{2}}$$

This number is either rational or irrational. If it's rational, then there *do* exist irrational numbers c and d such that c^d is rational, namely

$$c = \sqrt{2} \quad \text{and} \quad d = \sqrt{2}.$$

---

[1] For a proof of this, see the website.

On the other hand, if $\sqrt{2}^{\sqrt{2}}$ is not rational, then it's irrational. In this case, let

$$c = \sqrt{2}^{\sqrt{2}} \quad \text{and} \quad d = \sqrt{2}.$$

These are both irrational, and

$$
\begin{aligned}
c^d &= \left(\sqrt{2}^{\sqrt{2}}\right)^{\sqrt{2}} \\
&= \sqrt{2}^{\sqrt{2} \cdot \sqrt{2}} \\
&= \sqrt{2}^{2} \\
&= 2.
\end{aligned}
$$

Once again we have irrationals c and d such that c^d is rational.

That's it. If $\sqrt{2}^{\sqrt{2}}$ is rational, we have a c and a d. And if $\sqrt{2}^{\sqrt{2}}$ is irrational, we have a c and a d. Since $\sqrt{2}^{\sqrt{2}}$ is either rational or irrational, we've proved the theorem.

Constructivists reject this proof. They insist that a proof of the theorem must construct an irrational c and an irrational d such that c^d is rational. At the end of the proof we don't have explicit irrational numbers c and d such that c^d is rational.[2]

But what do *you* think? Is the reasoning here dangerous and unlawful? Or is it cute?

I can't resist telling a story about my son at age seven. The three of us, Freddy, my wife, and I, were eating dinner. Freddy was discussing "remainders." He defined a remainder as someone who was different in some respect from everybody else. He said "All of us are remainders. Papa, you're a remainder because you are eating a cheeseburger and Mama and I are eating hamburgers. I'm a remainder because I'm drinking milk and you and Mama are drinking cranberry juice. And Mama, you're a remainder because you're the only one who's not a remainder."

I'm not a constructivist. But I've always wondered, in what respect was Mama different?

Ethics is difficult in any field. It's time to return to creativity.

---

[2] For a while, this was the only proof of the theorem. Later, a constructive proof was found.

22

FUSION

SOLVING PROBLEMS is just one thing mathematicians and chefs do. A more important activity for both is *creating*. Chefs create new and sometimes wonderful dishes. Mathematicians create new and sometimes fascinating mathematical structures.

One might expect that there would be essential differences in how mathematicians and chefs do their creating. But, like problem-solving, there are common patterns. One of these is fusion. Bringing together two things that are old can sometimes get you something that's new.

FUSION CUISINE

Cuisines have been fusing since the dawn of time. Whenever two cultures meet, flavors, ingredients, techniques are exchanged. This happens casually and naturally.

Sometimes the result is more exciting to one side than the other. The cuisines created by immigrant groups in the United States (Chinese, Italian, Mexican, etc.) were bland variants of the originals, but they greatly enriched American palates.

Sometimes the result is greater than the sum of the parts. Mexican cuisine is a good example. It is a noble fusion of native American and Spanish cooking.

I like the idea of taking the technology of cuisine A and marrying it with the flavors of cuisine B. I've most often seen this where A is a European culture and B is an Asian culture. But I have some ideas for turning that around. Here's one of them. It marries Chinese technology with Italian flavors. There should be some matches here because both cuisines love noodles.

Potstickers Rustica[1]

For the dough:

2 1/2 cups flour
1 1/4 cups boiling water

Pour the water into the flour and stir rapidly. When stirring ceases to help, knead the dough, adding additional flour when required. Be careful, though, it's hot. Knead until the dough is smooth. Wrap it in plastic and let it rest a few minutes.

For the filling:

2 egg yolks
3/4 lb whole-milk ricotta
1/4 lb each diced prosciutto, mortadella, and mozzarella
2 Tb freshly grated Parmesan cheese
freshly ground black pepper to taste

For the cooking:

cooking oil
hot water
a frying pan with a cover

Beat the yolks, beat in the ricotta, then add the rest.

Take a fifth of the dough and form a log. Divide this in six pieces. Roll each piece into a thin disk, about 3–4 inches in diameter. Place a tablespoon or two of the filling on the disk, bring two opposite sides of the disk together over the filling and seal. Continue in this way with the rest of the dough.

Heat a frying pan with a thin layer of oil. When hot, place six dumplings in the pan. Cook, uncovered, until the bottoms start to brown (this should happen pretty quickly). Then pour into the pan a few tablespoons of water and cover the pan. The water will boil, steaming the filling and getting absorbed by the dough.

After about two minutes, uncover the pan. Continue cooking until the potstickers have lovely brown bottoms. Remove them, add more oil, and continue with the rest of the dumplings.

continued

[1] Adapted from Irene Kuo, *The Key to Chinese Cooking* (Knopf, 1977), and Marcella Hazan, *More Classic Italian Cooking* (Knopf, 1978).

---

continued

Notes:

- A potsticker is like a large ravioli that got fried. It seems to fit in the Italian line, as long as the filling is proportional to the dough, which this is.
- Of course, you don't actually want the dumplings to stick to the pan. A nonstick pan is helpful here, and I use one. But potstickers are older than Teflon, and a good ordinary frying pan should work.
- I keep a pitcher of hot water near the stove. Occasionally, I add too much water, more than the dough can readily absorb. Then I have to boil off the excess.
- Many other fillings work well here. I've done crab. I've done lobster. I've done smoked duck.

I spent a few very happy years in the Philippines. I am especially fond of one *merienda* (snack) dish, *palitaw*. It's a hand-made rice noodle, a noodle in the form of a thin, elongated disk. It is served sprinkled with sugar and toasted sesame seeds. It's tasty and chewy.

When I was looking for gluten-free pasta (see chapter 1), I wondered if *palitaw* might be useful. If you like the following dish, then it is useful. It could be described as a marriage of Philippine, Italian, and Jewish cuisines.

Loxitaw[2]

(For 8–10, as a first course. The servings will be small, but rich.)

4 Tb butter
3/4 cup heavy cream
8 oz lox or thinly sliced cold-smoked salmon
freshly ground black pepper
2 cups sticky (glutinous) rice powder or more
1 cup water
lemon slices

---

[2] Adapted from Efrem Funghi Calingaert and Jacquelyn Days Serwer, *Pasta and Rice Italian Style* (Penguin, 1987) and Enriqueta David-Perez, *Recipes of the Philippines* (National Book Store, 1972).

Melt the butter, then add the cream, then cook until thickened. Turn off the heat.

Mix 1 cup of water with 2 cups of sticky rice powder. Probably you will have a gooey mess. Add additional rice powder little by little until the dough has just enough cohesion to be rolled into balls between your palms. This is surprising stuff. It will be a little damp. It may remind you of silly putty.

Cut the slices of salmon into pieces roughly the size of silver dollars.

Put a pot of water on to boil. When the water starts boiling, take a bit of the sticky rice, roll it into a ball about 3/4 inch in diameter. With your fingers, press the ball into a thin, elongated disk. Drop the disk into the water. Continue making disks and dropping them in the water. You will need to dust your fingers with rice flour. After a while, the boiling disks will rise to the surface. When that happens, they're done. Scoop them out, drain them, and toss them into the sauce. Stir the sauce from time to time so the disks don't stick to each other.

This process will take a while. Add the salmon to the sauce at the end, stirring so that everything is nicely coated. Grind pepper over it and serve with the lemon slices.

MATHEMATICAL FUSION

Fusion happens all the time in mathematics. A famous example is calculus. The original problems of calculus were thought of as geometric: calculating areas, constructing tangents. In the seventeenth century, algebra was applied to those problems to solve them. Not everyone was happy about this—doesn't that sound like food cuisines? Some thought the use of algebra was a blemish on the purity of geometry (like using a blender to make pesto instead of pounding the ingredients in a mortar).[3] You would think that a method that succeeds would be universally approved, but mathematicians care about aesthetics too.

The field of topology, which came into being in the nineteenth century, is a complete fusion. It began with geometrical objects but ignored distance and angle and studied the objects with analysis.

---

[3] Thomas Hobbes provides a good example. He wasn't a mathematician so much as a philosopher and an influential critic. He described John Wallis's work on conic sections as "a scab of symbols."

Later, topology brought in algebra, set theory, and graph theory. Today, topology touches almost every branch of mathematics.

Let me tell you about how fusion is working with the boxes Fred and I introduced in chapters 10 and 18.

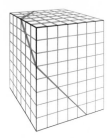

The inspiration for working with boxes came from rectangles (chapters 6 and 8).

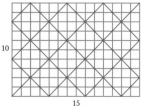

This is all geometry, but the rectangles touch on number theory. If the sides of the rectangle have a common factor (i.e., they are not relatively prime), then you can find a loop, a closed path that doesn't hit a corner.

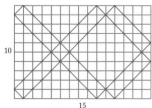

And if they don't have a common factor (relatively prime), every diagonal

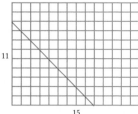

leads inevitably to a corner.

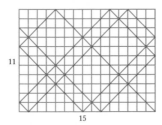

So what about boxes? It's true that if the dimensions have a common factor, there's a loop.

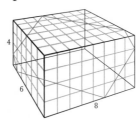

But even if they don't have a common factor, there could be a loop.

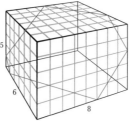

So what's going on? Fred and I decided to fuse number theory with geometry. We decided that maybe this was a different sort of "relatively prime." That is, maybe we could define "relatively box prime."

> **Definition** Three positive integers, a, b, c, are **relatively box prime** if the $a \times b \times c$ box doesn't have a loop.

The idea is that maybe "relatively box prime" would resemble "relatively prime" and that this would tell us something about boxes.

Well, we've had some success. Here's an example. In number theory, there's a simple way to find out if two numbers are relatively prime. What you do is take the two numbers

$$42 \qquad 33$$

and subtract the smaller one from the larger one.

$$
\begin{array}{ll}
42 & 33 \\
9 & 33 \qquad (42 - 33 = 9)
\end{array}
$$

You keep going, always subtracting the larger from the smaller,

$$
\begin{array}{ll}
42 & 33 \\
9 & 33 \qquad (42 - 33 = 9) \\
9 & 24 \qquad (33 - 9 = 24) \\
9 & 15 \qquad (24 - 9 = 15)
\end{array}
$$

and you don't stop until subtraction no longer changes things.

$$
\begin{array}{ll}
42 & 33 \\
9 & 33 \\
9 & 24 \\
9 & 15 \\
9 & 6 \\
3 & 6 \\
3 & 3 \\
3 & 0 \\
3 & 0 \\
3 & 0
\end{array}
$$

Then if you end up with 1 and 0, the numbers you started with are relatively prime, otherwise, they're not. In this case, they're not because you end up with 3 and 0 (3 is actually the greatest common factor of 42 and 33).

Another example: 42 and 31, which are relatively prime.

$$
\begin{array}{ll}
42 & 31 \\
9 & 31 \\
9 & 22 \\
9 & 13 \\
9 & 4 \\
5 & 4 \\
1 & 4 \\
1 & 3 \\
1 & 2 \\
1 & 1 \\
1 & 0
\end{array}
$$

This method is called the Euclidean algorithm.

Now what about boxes and the three dimensions? It took us a while, but I think we've found out what to do. Given three numbers, subtract the two smaller numbers from the largest number. If the dimensions of the box are 9, 26, and 15, for example,

$$
\begin{array}{ccc}
9 & 26 & 15 \\
9 & 2 & 15 \quad (26 - 15 - 9 = 2)
\end{array}
$$

And you keep going.

$$
\begin{array}{ccc}
9 & 26 & 15 \\
9 & 2 & 15 \quad (26 - 15 - 9 = 2) \\
9 & 2 & 4 \quad\ (15 - 9 - 2) = 4)
\end{array}
$$

As with the Euclidean algorithm, you keep going until things stop changing.

$$
\begin{array}{ccc}
9 & 26 & 15 \\
9 & 2 & 15 \\
9 & 2 & 4 \\
3 & 2 & 4 \\
3 & 2 & -1
\end{array}
$$

(And you keep going even if a number turns out negative!)

$$
\begin{array}{ccc}
9 & 26 & 15 \\
9 & 2 & 15 \quad (26 - 15 - 9 = 2) \\
9 & 2 & 4 \quad\ \ (9 - 2 - 4 = 3) \\
3 & 2 & 4 \\
3 & 2 & -1 \\
2 & 2 & -1 \\
1 & 2 & -1 \\
1 & 2 & -1 \\
1 & 2 & -1
\end{array}
$$

The negative number looks ridiculous here, but hang on. We can prove[4] that if you end up with

$$1, 0, \text{ and } 0,$$

---

[4] But I won't prove it here. There's a link on the website to an up-to-date discussion of the box.

or

$$1, 1, \text{ and } -1,$$

or

$$1, 2, \text{ and } -1,$$

then the original trio (9, 26, 15) are relatively box prime, that is, there are no loops. Otherwise, there are loops. For example, in the 9 × 26 × 15 box, any diagonal line

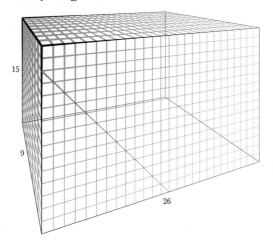

eventually hits a corner.

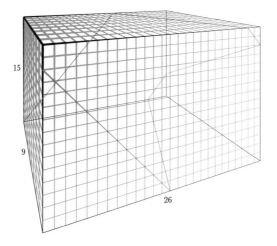

Notice that for the 5 × 6 × 8 box, this method

5	6	8
5	6	−3
5	4	−3
4	4	−3
3	4	−3

shows that there are loops (as we saw a few pages ago).

This is an odd fusion of number theory and geometry! There are many questions to answer. Given the size of the box, what are the lengths of the paths? How many loops are there?

If you want to play with this, here's our system for working on paper. Let's say you want to explore the $5 \times 6 \times 8$ box. Then draw the bottom, a 5×6 rectangle.

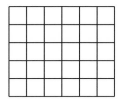

To investigate a path from a corner, start along the bottom.

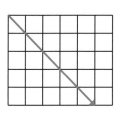

When you reach an edge, take eight steps (the height of the box) around the edge (you are climbing the sides of the box).

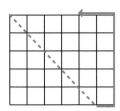

Now you're on top, so you keep going, crossing the top. . . .

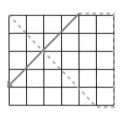 

And so on.

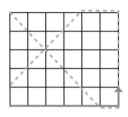 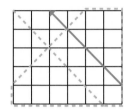 

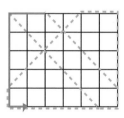 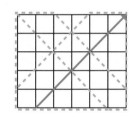 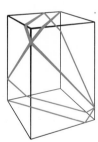

23

IT ALL COMES TOGETHER

MY GOAL in this book was to show that

MATHEMATICS AND GASTRONOMY ARE PRACTICALLY THE SAME.
In particular, I defended (successfully, don't you think?) the following claims:

- We do math and we cook for more or less the same reasons. (chapters 3, 4, 7, 8)
- We solve math problems the way we solve kitchen problems. (chapters 2, 6, 12, 13)
- We judge mathematics and food with much the same criteria. (chapters 5, 14, 17, 18, 20)
- Life in mathematics and life in gastronomy are remarkably similar. (chapters 1, 9, 10, 12, 15, 16, 19, 21, 22)

The bottom line is: if you're a successful, creative cook, you can do math. And if you're a successful, creative mathematician, you can cook. It's just a question of desire. The same attitudes, the same mental approaches, the same problem-solving skills propel you forward.

Now, a jaundiced reader might say,

Well, yes, but why cooking? Couldn't you have linked mathematics with literature? Or plumbing? Investment banking? Cinematography? Identity theft? What about math and short-track speed skating?

I have to say that the jaundiced reader is right.

Indeed, the larger message of this book is that mathematics is not so special. It is fundamentally similar to most other fields of endeavor. We thank the jaundiced reader for the observation and we apply it to the second point above, to get,

IF YOU CAN DO MATH, YOU CAN DO ANYTHING.

My stated goal was to show that math and cooking were (almost) the same. My *secret* goal was to elevate the status of fun. I think fun is great! Too many people think it's just . . . fun.

My argument for fun is pretty simple.

- If you're having fun doing something, you'll keep doing it. (chapters 1, 5, 13, 17)
- If you keep doing something, you'll get better at it. (chapters 6, 8, 12, 13, 17)

And these points lead us to:

FUN IS PRODUCTIVE.

This is hard to argue with. But a skeptical reader might say,

> You get better at it? Better at what? Most of what happens in this book is fluff—games, doodles, puzzles, card tricks! How is that "productive"?

Fluff? Fluff?

Well, maybe.

But when you say "fluff," I think you're talking about "fun stuff."

All our examples were (I hope) fun stuff. And that means that all the skills, all the power, all the smarts that you get from math you can get from fun stuff. In other words, with the help of the skeptical and jaundiced readers we've really shown

IF YOU CAN DO FUN MATH STUFF, YOU CAN DO ANYTHING.

And that says it all.

24

IT ALL FALLS APART

DID I say my approach, which might be described as

- Jump in and try something.
- If that doesn't work, try something else.
- Make mistakes and learn from them.
- Planning is for wimps.
- If you're having a good time, who cares if you get an answer?

always works? Well, it doesn't. I might as well tell you. You would probably find out anyway.

If the approach doesn't work for you, I want to say:

1. If you spend hours/days/weeks/years using my method on a problem without success, you haven't wasted your time because
   - You had fun (I hope).
   - You grew intellectually.
   - You now understand the problem at a deeper level.

2. You can always try standard methods. Read a book. Look at a recipe. Ask around. Go online. Your problem may be so hard that no one has been able to solve it.

3. But the importance of my method is that
   - It works a lot of the time.
   - Anyone can use it, regardless of talent or experience.
   - Its simple existence means no one who hasn't used it can justifiably say, "There's no point in trying. I can't do it."

To put it another way:

If you think the problem is a lack of smartness, you haven't tried dumbness!

4. And in fruitless attempts at problems, I'm way ahead of you. I've been banging my head against walls for centuries, possibly decades. Successful mathematicians and cooks are inured to failure. I solve perhaps 5 to 10 percent of the problems I tackle. That may be typical of mathematicians.[1]

I'll give you two illustrative examples, one in mathematics and one in cooking. In both I report failure—partial progress but no solution.

A SNEAKY, SNAKEY, STICKY PROBLEM

I saw an article about this problem in the *American Mathematical Monthly*, the most widely read math journal in the world.[2] Call it the *lamp problem*. In the lamp problem, n lamps are arranged in a circle, all lit, an arrow pointing to one lamp.

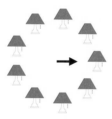

Pull the cord on this lamp if the lamp to its left (going counterclockwise) is lit. In this case, it is lit, so you pull the cord. Then move the arrow one lamp clockwise.

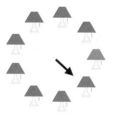

Now you repeat the operation. Since the lamp to the left is off, you don't pull the cord, but you do move the arrow.

[1] Cooks do better, I think. But it all depends on what you call a failure.

[2] Laurent Bartholdi, "Lamps, Factorizations, and Finite Fields," *American Mathematical Monthly* 187, no. 5 (2000), pp. 429–436.

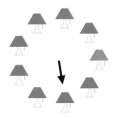

Continuing, this time you do pull the cord.

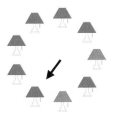

Just keep going. Eventually (maybe after a long time), all the lamps are back on. The lamp problem is the problem of finding out, given n (the number of lamps), how many steps it will take before the lamps are again all on. In the case of the nine lamps in the pictures above, it takes seventy-three steps.

The problem is unsolved. No one has found a formula for the number of steps.

There are many ways to approach a tough problem. One way is to chop it up into smaller problems. The article I read on the lamp problem did this and it solved a few of the smaller problems. But the problem as a whole remains unsolved.

Another approach is to see the problem as part of a bigger problem.

Sometimes the bigger problem is actually easier because you're looking at things from a different point of view. This is what I did. I found a bigger problem. But it wasn't easier. I call it the *snake problem*.

Consider a snake of length j ($j = 9$ in this case).

Suppose that it snakes around in a serpentine fashion:

Then breaks,

shifts,

and reforms as a new snake.

Basically, the operation rearranges the numbers from 1 to j.

The snake problem is the problem of finding, given j, how many times you have to snake around, break, and reform to return to the original snake. In the case of the snake with nine segments, it takes seven steps.

It turns out that if you can solve the snake problem you can solve the lamp problem! The lamp problem for n lamps is the same as the snake problem for snakes of length

$$j = 2^n.$$

For example, to figure out this set of lamps,

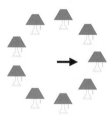

you only need to analyze this snake!

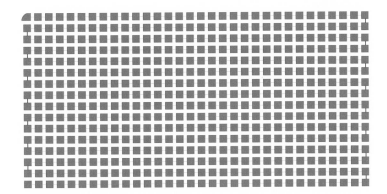

So the snake problem is a bigger problem. In fact, only part of the snake problem (snakes of length 1, 2, 4, 8, 16, . . .) is needed for the lamp problem.

But alas, I couldn't solve the snake problem any more than I could solve the lamp problem. Instead, I came up with a third problem. I call it the *problem of the sticky bounces*.

A "sticky bounce" is when a particle approaches a surface

and bounces at an angle of 90° to the surface.

Now consider an equilateral triangle with sides of length q (in this case $q = 7$).

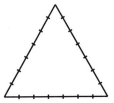

Start on the bottom, a distance of 1 from the lower left corner. Go straight up until you hit a side of the triangle. Bounce off in sticky bounce fashion, at a 90-degree angle from the side.

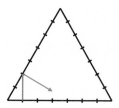

It turns out that if q, the length of the side, is odd, you will eventually return to where you started.

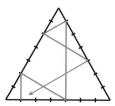

The problem of the sticky bounces is the problem of finding out, given q, how many segments there will be in the path before you return to your starting point. For the 7-triangle (the triangle with side 7) there are six segments.

Sticky bounces is actually off-track. The sticky bounce problem doesn't connect to snakes of length 2, 4, 8, 16. . . . What I've found is that if, for a triangle of (odd) side q, the number of segments in the path is even, then that number is the answer for a snake of length $2q$. In other words, since the 7-triangle (above) has six segments, the 14-snake,

needs six steps.

But if the number of segments is odd, then the answer is half the answer for a snake of length $2q$. In other words, since the 11-triangle has five segments,

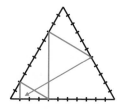

the 22-snake will need ten steps.

But alas! I can't solve the problem of the sticky bounces!

So, actually, what did I accomplish? Not much! But I had fun. And I invented a couple of problems that interest me and may interest others.

And I'm not done. I plan to work more on this. I'll post news about the problem on the book's website (look there now for sticky refractions).

A CELEBRATORY VEGETARIAN DISH

There are thousands of vegetarian recipes. Vegetarian cooks are tremendously creative. The dishes are wonderfully flavorful and nutritious. They don't satisfy every desire, of course, but they are versatile and varied.

There are vegetarian dishes for every occasion—except one. There is nothing to take the place of roast turkey (with the trimmings). There isn't the monumental celebratory vegetarian dish.

I don't mean a vegetarian composition that looks like or tastes like roast turkey (Tofurkey, for example). I'm thinking of turkey as an institution. Turkey is more than a meal. It's a celebration of plenty, a celebration of family,[3] a celebration of country. It's a multi-dimensional feast. It takes hours to prepare, hours to cook, hours to eat, and weeks to eliminate.

There are some impressive vegetarian creations—Molly Katzen's Enchanted Broccoli Forest, for example. But I haven't seen anything that truly takes the place of roast turkey. The challenge I issue here (and that I have taken up myself repeatedly) is to create a vegetarian dish that says:

1. We're a great family, and
2. We live in a great country, and

---

[3] Unless your family includes a vegetarian.

3. Somebody who loves me very much spent hours in the kitchen preparing this amazing dish that I love to eat exactly once a year.

I've tried and I'll keep trying. I'm not a vegetarian, but I often cook for vegetarians, on Thanksgiving, for example.

Vegetable Wellington

This was an early attempt. I didn't record exactly what I did. But beef Wellington is a marvelous dish. A tenderloin is marinated then coated with mushroom duxelles, Madeira, shallots, and foie gras, and then baked in a puff pastry case.

I still think a vegetable Wellington might work. Whatever I did was appreciated and enjoyed, but I was disappointed and I didn't try it again.

Stuffed Eggplant on a Spit

This shows the dangers of attempting something halfway. I stuffed it. I put it on the spit. I set it over the fire. All seemed well. It took on color. It puffed.

I went back to the kitchen to prepare sauces. I think I will never forget the shrieks from my guests when the whole thing fell into the flames.

Stuffed Dragonskin

I thought this had great potential. "Dragonskin" is tofu skin, ultrathin sheets of tofu. It comes frozen. You thaw the sheets, paint them with oil and something flavorful, then wrap them in many layers around something especially flavorful. You fry this. The outer layers become crisp.

I wrapped traditional turkey stuffing. The idea was that the tofu skin would provide the protein in the meal. I could accompany the dish with cranberry sauce, mashed potatoes, and some sort of gravy.

It wasn't bad. But it lacked heft. Turkey meat is a solid presence. This didn't have that presence.

Vegetarian Cassoulet

Cassoulet is a bean dish enriched with meats and sausages. Beans are central, but the meat is quite important: pork, lamb, and goose. I've had vegetarian cassoulet that was not impressive. I was determined to do something special.

I pulled out all the stops. I took as a basis Julia Child's recipe for cassoulet in *Mastering the Art of French Cooking.* The result was very good. But it fell short of the goal. Here's what I did:

2 lb cannellini beans

I dropped those in boiling water. I brought the water back to a boil, then turned off the heat and let the beans soak for an hour.

3 small golden beets
2 parsnips
1 fat carrot
2 big shallots
some baby bella mushrooms
2 tomatoes
1 head of garlic
Those I coated with peanut oil and then roasted.

1 recipe of shamburger (p. 46)
allspice
bay leaf
brandy
garlic
truffle (okay, truffle salt)

The herbs are from Julia's recipe for the sausage and I added them to the shamburger. I formed the mix into balls, browned them, and deglazed the pan with something, I forget what.

lots of chestnuts

Those I boiled, then shelled, then cooked in butter. I had about 1 1/3 cups when I was done.

1 sweet red pepper

I blackened that over a gas flame, flaked off the burnt skin, chopped up the flesh and cooked that in butter.

continued

continued

1 large onion
I caramelized that in butter.

1 package Classic French Essence de Champignon Gold[4]
2 cups white vermouth
2 bay leaves
1 sprig thyme
some pieces of Parmesan
the beans
I cooked all that for an hour and a half.

some chunks of cheddar
some pats of butter
Finally, I put these and everything else together in a casserole and baked it for a while.

What did I accomplish? Not a whole lot! But I had fun. And I dreamed up a few dishes that interest me and may interest others.

And I'm not done. I plan to work more on this. I'll post news on the book's website (look there now for tofu napoleons).

---

[4] A brand-name concentrated base for mushroom broth.

25

A PROOF AND A PUDDING

I'M GOING to close this book with a proof and a pudding. Both are quick. Both use surprisingly few ingredients. Each is, in its own way, sweet and a bit mysterious.

A PROOF

When I was in college I decided, for the fun of it, to come up with a mathematical definition of "nice set."

What should be true about a set that's *nice*? Well, I thought, a nice set shouldn't have anything bad in it. Its proper subsets, for example, shouldn't be bad. A proper subset is a subset which isn't the whole set. The set $\{a, b\}$, for example, has four subsets: $\{\ \}$, $\{a\}$, $\{b\}$, $\{a, b\}$, but only three of them are *proper* subsets—the last subset isn't proper because it's the whole set.

So the proper subsets shouldn't be bad. Maybe that means they should be nice! That idea led to this definition:

> Definition: A set is *nice* if and only if all its proper subsets are nice.

But that's a ridiculous definition! I'm defining "nice" in terms of "nice"! How can that make any sense?

But ridiculous or not, this definition actually produces results.

> Theorem: All finite sets are nice.

> Proof: Suppose that the theorem is false. We'll see that this leads to a contradiction, an impossibility. This will show us that the theorem can't be false and must be true.

So, suppose the theorem is false. Then there must be finite sets that aren't nice. Let Z be the smallest finite set that isn't nice. Since Z isn't nice, it must have a proper subset, W , that isn't nice. Since Z is finite, W is smaller than Z (remember W doesn't equal Z).

But we chose Z to be the *smallest* set that isn't nice! That means that all smaller sets are nice! So W is nice. That's a contradiction!

So the theorem *can't* be false. So it must be true!

Q.E.D.

A PUDDING

Imagine a pudding with just three ingredients. You put them together and leave them alone. You don't cook them. You don't watch them. You don't give them any instructions or advice. But somehow they know what to do. They jell and make puddings.

Blueberry Mousse

1 pint blueberries
1 large lemon
3 Tb sugar

Wash and dry the blueberries and remove any stems. Place the blueberries in a blender. Add the grated rind of the lemon and its juice. Add the sugar.

Blend.

Pour the blended ingredients into six goblets or custard cups. Cover them with plastic wrap and chill them in the refrigerator.

That's it.

A proof and a pudding.

Both are simple. And both leave you wondering why they work.

INDEX